Circulation and Gover
Asian Medicine

This book unpacks the organized sets of practices that govern contemporary Asian medicine, from production of medications in the lab to their circulation within circuits and networks of all kinds, and examines the plurality of actors involved in such governance.

Chapters analyze the process of industrialization and commercialization of Asian medicine and the ways in which the expansion of the market in Asian medicines has contributed to the inscription of products within a large system of governance, greatly dominated by global actors and the biomedical hegemony. At the same time, the contributors argue that local actors continue to play a major role in reshaping the regulations and their implementation, thus complexifying the trajectory of the remedies and their natures. Examining in particular the plurality of actors involved in governance and circulation, and the converging or conflicting logics actors follow in regard to negotiations and tensions that arise, the book brings a unique multi-layered contribution to the study of governance and circulation of Asian medicines, offering further proof of their fluidity and resilience.

Filling a significant gap in the market by addressing circulation and governance of Asian medicines in Asian countries, including Bangladesh, Myanmar, and Singapore, this book will be of interest to students and scholars in the field of Asian studies, Asian culture and society, global health, Asian medicine, and medical anthropology.

Céline Coderey is an Anthropologist and Research Fellow at the Asia Research Institute of the National University of Singapore.

Laurent Pordié is an Anthropologist and Senior Researcher with the National Center for Scientific Research (CNRS-CERMES3), France.

Routledge Contemporary Asia Series

Political Participation in Asia
Defining and Deploying Political Space
Edited by Eva Hansson and Meredith L. Weiss

Religious and Ethnic Revival in a Chinese Minority
The Bai People of Southwest China
Liang Yongjia

Protecting the Weak in East Asia
Framing, Mobilisation and Institutionalism
Edited by Iwo Amelung, Moritz Bälz, Heike Holbig, Matthias Schumann and Cornelia Storz

Middle Class, Civil Society and Democracy in Asia
Edited by Hsin-Huang Michael Hsiao

Conflict in India and China's Contested Borderlands
A Comparative Study
Kunal Mukherjee

Transcontinental Silk Road Strategies
Comparing China, Japan and South Korea in Uzbekistan
Timur Dadabev

Sino-Pakistani Relations
Politics, Military and Regional Dynamics
Filippo Boni

Circulation and Governance of Asian Medicine
Edited by Céline Coderey and Laurent Pordié

For more information about this series, please visit: www.routledge.com/Routledge-Contemporary-Asia-Series/book-series/SE0794

Circulation and Governance of Asian Medicine

**Edited by Céline Coderey
and Laurent Pordié**

Routledge
Taylor & Francis Group

LONDON AND NEW YORK

First published 2020
by Routledge
2 Park Square, Milton Park, Abingdon, Oxon OX14 4RN

and by Routledge
605 Third Avenue, New York, NY 10017

First issued in paperback 2022

Routledge is an imprint of the Taylor & Francis Group, an informa business

British Library Cataloguing-in-Publication Data
A catalogue record for this book is available from the British Library

Library of Congress Cataloging-in-Publication Data
Names: Coderey, Céline, editor. | Pordié, Laurent, editor.
Title: Circulation and governance of Asian medicine / edited by Céline Coderey and Laurent Pordié.
Other titles: Routledge contemporary Asia series.
Description: Abingdon, Oxon ; New York, NY : Routledge, 2020. | Series: Routledge contemporary Asia series | Includes bibliographical references and index.
Identifiers: LCCN 2019032570 (print) | ISBN 9780367225346 (hardback) | ISBN 9780429275418 (ebook) | ISBN 9781000650730 (adobe pdf) | ISBN 9781000650822 (mobi) | ISBN 9781000650914 (epub)
Subjects: MESH: Pharmacognosy—legislation & jurisprudence | Legislation, Drug | Medicine, Traditional—standards | Drug Industry—economics | Asia
Classification: LCC RS179 (print) | LCC RS179 (ebook) | NLM QV 733 JA1 | DDC 615.3/21095—dc23
LC record available at https://lccn.loc.gov/2019032570
LC ebook record available at https://lccn.loc.gov/2019032571

ISBN: 978-1-03-240122-5 (pbk)
ISBN: 978-0-367-22534-6 (hbk)
ISBN: 978-0-429-27541-8 (ebk)

DOI: 10.4324/9780429275418

Typeset in Times New Roman
by Apex CoVantage, LLC

Contents

Table

Contributors

Arielle A. Smith completed her doctorate and postgraduate teaching in medical anthropology at the University of Oxford (2004–2010). She subsequently taught at the University of Montana (2011–2012) and as traveling faculty for International Honors Program/ SIT (2012–2014). Most recently, she conducted postdoctoral research at CERMES3 (a joint unit of CNRS, EHESS and Inserm) in Paris (2015) and preliminary research on tribal health/ healing in the United States (2016–2017), and published her first book on the topic of Chinese medicine in Singapore (2018).

Caroline Meier zu Biesen is a Research Associate at the Institute of Social and Cultural Anthropology (Freie Universität Berlin) and a Research Fellow within the ERC-project on GLOBHEALTH (CERMES3, Paris). Her research interests focus on global health governance, social inequality and health, transnational drug circulation, HIV/AIDS, malaria, and traditional medicine. She has conducted long-term fieldwork in Eastern Africa and India. Her current DFG-funded research project focuses on the collaboration among traditional and biomedical practitioners in the management of noncommunicable diseases (NCDs) in Zanzibar.

Céline Coderey is a Medical Anthropologist currently appointed as Research Fellow at the Asia Research Institute of the National University of Singapore and as a Teaching Fellow in Tembusu College. Her research covers several aspects of the "therapeutic field" in contemporary Myanmar: the institutionalization of traditional medicine, the governance and circulation of medical products, (the obstacles to) the accessibility of biomedical healthcare services, notably in the sector of HIV and mental health, and practices of divination and alchemy. Her current projects looks at how the political and social transformation within the country affect both healers' practices and patients' health-seeking process.

Eunjeong Ma is a collegiate associate professor in the Department of Creative IT Engineering at Pohang University of Science and Technology (POSTECH), Pohang, South Korea. Having been trained in Science and Technology Studies, she works on the sociocultural dimensions of pharmaceuticals and medical technologies, along with engineering culture and education.

Karen M. McNamara recently completed a joint Postdoctoral Research Fellowship in the Science, Technology, and Society cluster of the ARI and a Teaching Fellowship at Tembusu College, National University of Singapore. She has conducted anthropological research on traditional medicine, neoliberal governance and care, and migration and health in Bangladesh, India, and Singapore. Her publications have examined the intersection of traditional medicine, herbal remedies, and the emergence of a pharmaceutical industry in Bangladesh, including her chapter, "Medicinal Plants in Bangladesh: Planting Seeds of Care in the Weeds of Neoliberalism," in *Plants & Health: New Perspectives on the Health-Environment-Plant Nexus* (Springer, 2016). Her current research examines the politics and possibilities of care in the medical travels of Bangladeshi patients to India and Singapore.

Laurent Pordié is an anthropologist, Senior Researcher with the French National Center for Scientific Research (CNRS) at the CERMES3 in Paris. Among his publications are *Tibetan Medicine in the Contemporary World* (Routledge, 2008) and *Les nouveaux guérisseurs* (EHESS, 2013).

Liz P.Y. Chee is Research Fellow at the Asia Research Institute (ARI) and Fellow at Tembusu College, both at the National University of Singapore (NUS). She is the first graduate of the NUS-Edinburgh joint PhD program. She is an historian, and her main research is on the use of animals in Chinese medicine and foodways.

Simeng Wang is a Permanent Research Fellow at CNRS and faculty member at the CERMES3 (Research Centre, Medicine, Science, Health, Mental Health and Society). She earned a PhD in Sociology from the École Normale Supérieure in Paris. Since 2009, she has been working on Chinese immigration in France, first about mental health issues, and since then, through other angles: mobility, transnationalism, political participation, etc. She is the author of several books and scientific articles dedicated to Chinese diasporas, to health and mental health issues and to contemporary China (see among others: Simeng Wang. 2017. *Illusions and suffering. Chinese migrants in Paris*. Paris: Presses of École Normale Supérieure. 220p. Simeng Wang [with Isabelle Coutant, eds.]. 2018. *Mental Health and mental suffering: a research subject for social sciences*. Paris: CNRS Editions. 416p. Cf. https://cnrs.academia.edu/SimengWang). She is currently co-leading the social sciences interdisciplinary research "Emergence(s)" program granted by the City of Paris (2018–2021) "Chinese of France: Identifications and Identities in Transition." She is also an elected member of the executive committee of the French Sociology Association since 2017.

Acknowledgments

First of all, we wish to thank the Asia Research Institute (ARI) of the National University of Singapore which generously funded and hosted the conference "Governance and Circulation of Asian Medicines," which led to the advent of this book. We also wish to thank the EC-funded GLOBHEALTH Project at the CERMES3, Paris, for supporting the conference by funding the travels of our European guests. This conference facilitated the gathering of numerous scholars interested in reflecting on the question of governance and circulation of Asian medicines.

Another special thanks goes to A/P Gregory Clancey, leader of the Science, Technology, and Society cluster at ARI, who not only enthusiastically welcomed the idea of a conference on the topic, but has also supported this project through all its stages, providing guidance as well as insightful comments on the texts, notably the introduction.

We also wish to thank friends and colleagues who have helped us throughout this project by providing general support and advice, notably Karen M. McNamara, Jeremy Fernando, and Ayo Wahlberg, and those who have helped us to refine our thoughts and improve our writing, whether through comments and suggestions, or through editing. In particular: Ms Nyx Chong for her reading of the chapters and for providing insightful comments on each of them; Laurent Pordié, Ayo Wahlberg, Stephen Campbell, Elliott Prasse Freeman, Gerard McCarthy, and Andrew Ong, for providing insightful comments on the introduction; Jeremy Fernando, Ranjan Circar, and Saharah Bte Abubakar, for editing some chapters of the book.

Finally, we are immensely indebted to all the healers, medicines vendors, company owners, medical authorities, government officials, staff of international or global organizations, and other informants we have worked with, and who have shared so generously with us their time and knowledge, thus providing us with the *materia prima* – the main ingredients – of this book.

Introduction

Governance and circulation of Asian medicines

Céline Coderey

In the last 20 years, the circulation of Asian medicines has experienced an unprecedented expansion at the national, regional, and global levels. Existing circuits have been intensified, and new ones have emerged. Urban Indians can now turn to any nearby pharmacy to purchase Ayurvedic products instead of going, as they used to do in the past, to a traditional healer who would compose an ad hoc remedy after having diagnosed the specific condition affecting them; Burmese and Filipino migrant workers based in Singapore can easily purchase their favorite branded remedies from their native countries in the big shopping malls of the city-state; and Americans in disaffection with biomedicine and fascinated by what they see as exotic treatments can now get Tibetan pills delivered to their doorstep with a simple click on the internet or buy them – sold under the label of food supplements or health products – in some local alimentary shops. If the intensification and expansion of circuits touches some medicines more than others, with Ayurvedic (Banerjee 2009; Bode 2002, 2008; Pordié 2014, 2015; Wujastik and Smith 2008), Chinese (Taylor 2005; Hsu 2009; Zhan 2009), and Tibetan medicines (Jane 2002; Pordié 2008, 2016; Craig 2012; Adams 2001, 2002; Saxer 2013) already representing a "global commodity," it is undeniable that many other medicines produced in the region are following a similar trend (Nguyen 2012 on Vietnam).

This phenomenon is the result of the increase in the demand for these products, combined with the expansion and liberalization of the market, both resulting from the intensification and acceleration of what Appadurai (1996) calls "global flows," i.e., the circulation of ideas, goods, technologies, financial wealth, and people around the globe commonly referred to as "globalization."

The high "circulability" of Asian medicines has been prepared – fostered and shaped – by the industrialization and mechanization of the manufacturing process and the commercialization of the products. Either initiated by corporate firms or encouraged by the states as part of national(istic) projects of institutionalization/valorization of traditional medicines, the industrialization and standardization of Asian medicines have entailed dramatic changes bearing on both medical epistemology and therapeutic practice. Medicines turned from compounded powders, balms, and decoction crafted by healers on the basis of ingredients personally collected in the wild and then dispensed (when not prepared) according to the specific condition of the patient, to ready-made pills, capsules, and balms, mechanically produced, packed in colorful boxes displaying the name and the logo of the brand, and distributed to a more or less extended network of shops where they are sold alongside biomedical drugs. This process of industrialization and commercialization has contributed to the separation of medicines from their original social context and their integration into new assemblages aggregating ideas, techniques, materials, and people (Serres and Latour 1995; Latour 2000; Rabinow 2003; Collier and Ong 2005), whereby medicines have become commercial commodities apt to be circulated in the region and beyond.

The transformation of Asian medicines did not, however, stop here. The expansion of the market of Asian medicines has contributed to the inscription of these products within a large system of governance which has actually evolved in parallel with the expansion of the circulation itself. This system is an assemblage made of both local and global actors – people, institutions, medicines, techniques, financial resources, knowledge, values, and interests – and crossed by significant power relations which are sources of significant frictions (Tsing 2004). This is all the more true given that, if governance ideally means a genuinely pluralist configuration of dialogue and shared objectives, in reality such homogeneity is seldom reached because different actors often operate according to specific values and agendas (be they political, economic, social, medical, or legal) that are not necessarily shared by the others. The inscription of Asian medicines within this system has come to complexify their nature and to multiply their meanings, in the same time that it has largely oriented or reoriented, sometimes favored or hindered, their circulation. It is this complex and dynamic intertwinement between circulation and governance of Asian medicine that this book will explore, thus opening up avenues for comparative studies on traditional medicine worldwide.

Toward a global regulation

Before the industrialization and commercialization of Asian medicines, their production and circulation were mainly regulated by the actors involved: healers, traders, and vendors, operating according to medical and economic logics. Yet the institutionalization of traditional medicines, commonly associated with their integration into the national health systems,[1] has come to inscribe these medicines within the state bureaucracies through the intervention of different bodies: ministries of health, ministries of finance, ministries of industry. If, at that stage, global actors were already influential– notably through the World Health Organization (WHO), which supported the institutionalization of traditional medicines – their role expanded when the products started to circulate far beyond the national borders. The number of international and global actors increased as well so as to include, in addition to the WHO, the World Trade Organization (WTO), the Association of Southeast Asian Nations (ASEAN), the World Intellectual Property Rights Organization (WIPO), the World Bank, and others.

The main role of these global actors is, in principle, to grant the international homogenization of regulations in order to ensure the smooth circulation of safe and quality products. To do so, they provide norms and guidelines which address the different stages of the social life of medicines (Whyte et al. 2002) from their conception to their consumption, and which touch at different nodes of the pharmaceutical nexus (Petryna and Kleinman 2006). The main norms and guidelines are drug laws (concerning licensing and registration), good manufacturing practices (GMP), good distribution practices (GDP), "border regimes" – administrative efforts aimed at regulating the transition of herbs and traders across borders (Saxer 2013, 109) – and the licenses for farmers, traders, manufacturers, sellers, and practitioners.

One of the aspects that has been the most strongly highlighted in the literature (Adams 2002; Saxer 2013; Craig 2012; Pordié 2008, 2010, 2011), is the major influence the biomedical paradigm has on the regulatory system as a consequence of the central role the WHO plays within that system. The WHO indeed represents one of the main actors of "global health governance" (WHO 2018) and acts as "the directing and co-ordinating authority on international health work." Now, most regulations and guidelines established by the WHO and the GMP in particular, are based on standardized models originally developed for the regulation of pharmaceutical biomedical drugs (Saxer 2013; Craig 2012). The application of biomedical regulations on Asian medicines suggests that they are what Collier and Ong call a "global form," characterized by its capacity for "decontextualization and recontextualization, abstractability and movement, across diverse social and cultural situations and spheres of life" (Collier and Ong 2005, 11). This transferability and applicability elsewhere are justified through the idea based on the false premise that science – the principle on which biomedicine stands – is universal, rational, objective, and neutral, and disconnected from culture, society, and politics. Yet, as Foucault (1979), Latour (1988, 1993) and others have shown, Western science and its paradigms are socially, culturally, and politically grounded, very much like any other form of knowledge, including medical knowledge. This is something that Craig (2012) has named "social ecology," an expression which refers to the interrelationships among environmental, socioeconomic, biological, political, and cosmological sources of, or explanations for, health problems. The obligation for traditional medicine to fit the biomedical model in order to be recognized and legitimized, and to be allowed a wider circulation within the global market, is a clear expression of the biomedical global hegemony and articulates what Foucault would call the biopolitics of pharmaceutical governance. In a similar vein, Craig and Adams (2009) speak of "global pharma" and define it as "the dominant discourse about efficacy, scientific legitimacy," as well as a "technique of governance."

This political game is likely to have – and actually does have – concrete implications for the formulation – and thus the essence – of traditional medicines, alongside their circulation. This is all the more so given that traditional medicines on the one side, and Western medicine on the other side, rely on epistemes and paradigms which are different and, according to many, often largely incommensurable (Kuhn 1970, 1982; Saxer 2013; Craig 2012). Although these incongruences are well known, we shall state here at least the main ones. Asian medicines are traditionally composed of a high number of ingredients, and, for some, may include doses of heavy metals, notably mercury, while GMP guidelines authorizes only a limited number of ingredients and generally proscribes heavy metals. They may also include the parts and tissue of animals listed by governments and non-governmental organizations (NGOs) as endangered. Moreover, in many Asian medicines, the efficacy of the products is not conceived to only stem from the intrinsic properties of the material ingredients, but also from other factors related to the moral nature of the person doing them, the ritual context, or the auspicious moment of the fabrication. Western science does not recognize these aspects. It qualifies them as "primitive/prescience/irrational" (Jagtenberg and Evans 2003, 323) and excludes them from efficacy, safety and quality testing procedures, at best relegating them to a status of "placebo" (Wahlberg 2008a).

The way biomedical governance operates is very subtle. Indeed, the embracing of biomedical regulations and GMP leads actors to think and behave differently than they are used to – they modify the way people see and fabricate medicines. This also implies the control of bodily gestures and movement within space. As stressed by Saxer (2013), they thus represent a form of governmentality (Foucault 1991) – the art and techniques by which citizen-subjects are governed. Some of these regulations go very far and take the form of discipline and self-discipline (Foucault 1979). They indeed inscribe new systems of thoughts and values within peoples' bodies through "technologies of the self": forms of knowledge and strategies that "permit individuals to effect by their own means or with the help of others a certain number of operations on their own bodies and souls, thoughts, conduct, and way of being" (Foucault 1988, 18). We are thus confronted with "regimes of governance [that] regulate people's behaviors through the application of political power, which becomes embodied in human experience, including making medicines" (Craig 2012, 168).

Local governance and the circulation of medicines across international borders

Despite their power, global actors have not erased the role of the state and of other local actors who have progressively multiplied to include donors, private enterprises, research centers, pharmaceutical companies, and others. More and more influential, these non-state actors operate sometimes in conjunction and sometimes in opposition to the state and the global actors. The state does maintain a certain agency and can decide to what extent and in which way to apply norms and guidelines established by global actors; it can also determine the very content of the different norms: GMP, drugs laws and border regimes. The way it does it, however, depends on the local specific sociopolitical-economic configurations, alternative understanding of what medical safety and efficacy mean, and manufacturers' market ambitions, as well as the influence of the other influential actors. Most regulations individual states implement are thus not ready-made forms, but local creations states produce by reshaping global forms depending on local characteristics (Saxer 2013). In other words, as global forms, GMP and other regulations are thus not only decontextualized from the original context and transplanted as such elsewhere, but are recontextualized in the new context; they are "reterritorialized" in a new "assemblage" (Collier and Ong 2005, 4). This localization corresponds to what Michael Burawoy et al. (2000) term as "grounding globalization," whereby, as phrased by Pordiè (2013, 8), "territory, land, and nations" are allowed to express themselves in globalization processes, to "play an instrumental role in shaping the global, in diversifying and complexifying it". In fact, "it is well-known that global processes are at all times local processes embedded in territories, communities, households, individuals and objects" (Ibid). As mentioned by Law (2004), "The global may then be seen as small and diversified rather than big and homogeneous" cited in Pordiè 2013, 8. Hence, the importance of taking the perspective of the locality, where we are standing, and examining how things unfold there (Wahlberg 2018, 10).

On the ground, this means that regulations among countries differ. GMP, drug laws, border regimes, and the licensing system in Myanmar differ from GMP, drug laws, border regimes, and the licensing system in India or in Korea. It also means that within any one country, regulations or the obligation to stick to these regulations might vary depending on the product and the destination the product is intended for. In Tibet, only commercial products distributed through the market need to comply with GMP, while those produced for the state hospitals do not (Saxer 2013); in Bangladesh and Myanmar, all products produced in the country are supposed to comply with GMP (McNamara, Chapter 1 of this volume; Coderey, Chapter 3 of this volume); in Korea, only standardized herbal medicines are allowed to circulate and to be granted that label they must be inspected through the same evaluation system applied to chemical drugs (Ma 2010). This also goes for the United States, where products have to go through laboratory tests to prove their medical properties in order to be labeled and sold as drugs, and the procedure is the same for herbal and chemical drugs.

Also contributing to the international and interregional difference in the governance system is the fact that Asian medicines – and thus their practitioners, medicines and techniques – are attributed a different position and recognized differently in the medical arena of respective countries, including from the legal point of view. In many Asian countries, traditional medicine is part of the official healthcare system, but there are cases where a medicine officially recognized is not included in national health, as is the case of Chinese medicine in Myanmar. In European countries and the United States, some Asian medicines may have a legal status under the guise of complementary alternative medicine (CAM), a label adopted elsewhere in Asia. CAM are subject to specific, relatively light, regulations which vary from country to country. The legal status and the category attributed to one medicine greatly impacts on its social recognition, the possibility for it to be covered by health insurance, and thus its attractiveness to potential consumers.

Finally, the way the production and circulation of Asian medicines are regulated depends on – and thus varies with – the fact that these medicines are recognized – and categorized – as medicines in the first place. This basically means that they are different from food – and this is far from obvious, given that the boundaries between food and medicine are often blurred and vary from country to country (Etkin and Ross 1982; Temkin 2002; Engelhardt 2001).

This diversity is not only interesting in that it shows the weight of locality within the governance system, but also because it has very practical implications when products start to circulate outside national borders and aim for several destinations. Depending on the region or the country where a product is marketed, manufacturers may or may not have to comply with different regulations. For instance, although circulation within the Asian region is relatively smooth, it is much harder for Asian products to penetrate the Australian, European, Canadian, and American markets.

Moreover, even when products do not travel, ideas and values do so – and this is also true for ideas of modernity and science and for biomedical etiologies. Western conceptions of medicine and biomedical etiologies reached the East much earlier than the introduction of GMP and other regulations; thus creating new needs and demands from consumers that global governance then just came to reinforce (Pordié 2008, 2013). For many Asian countries, this adoption of Western and biomedical ideas was related to the post-colonial push for the modernization of the newly emerging nation-states as China (Taylor 2005), Korea (Ma 2010), Vietnam (Wahlberg 2014), Singapore (Smith, Chapter 4 of this volume), and Myanmar (Coderey, Chapter 3 of this volume).

Therefore, whether intended for local consumption or for export, medical products have to adjust or be reinvented to comply with the expectations of global consumers and regulators.

Fluidity and resilience of Asian medicines

As a response to this regulatory framework and its variations, medicines all across Asia are changing, whether in their design and content, and in the manufacturing and production process (simplifications, innovations); in their labels and marketing strategies; in terms of their prescribers, providers, and consumers; or in each of these ways at once. Significantly, new forms of medicines – or at least new denominations and categories – have emerged, notably "herbal medicines," "herbal medicinal products," "health functional foods," and "health products," strategies created and adopted to make the circulation of certain products smoother. Herbal medicines ground their legitimacy in their apparent similarity to biomedicine and their "natural" appeal, while health functional food/health products rely on their liminal position between food and medicine, an important characteristic of Asian medicines. Equally significant and related to the latter is the emerging necessity to protect one's materials or products through intellectual property laws and patenting systems.

If many actions are accomplished in compliance with the laws, others are rather circumventing strategies which profit from the gray zones opened by the blurredness or the cross-national inconsistencies of regulations, as well as from the mobility of products and individuals. All these actions are part of the governance system very much like the regulations coming from the state and global bodies, as already highlighted by Quet et al. (2018, 2–3). Interestingly, as noted by Saxer (2013, 134) in relation to border regimes, the breaking of rules is crucial for their survival, for making them work or appear to work. Saxer also suggests that "regime connotes state power, while border is liminal space where state power is potentially vulnerable. Border regimes are more flexible, manipulable and prone to change than related apparatus and legal frameworks on which they are based" (2013, 110). We can make the assumption that this could be valid for the other rules, as well, and that it is the balance between respecting rules and breaking rules – allowed by the blurred space provided by the system of governance – that grants the viability of a medicine's market. This hypothesis recalls and converges with the Foucauldian idea about security that an accepted margin of deviance is necessary in order to allow the system to be viable (Foucault 1979).

The fact that medicines are transformed alongside their trajectory, acquiring new names, labels, and legal or illegal status, reveals how much medicines are mobile, fluid, and adaptable objects whose nature, identity, and meaning are largely affected by regulation regimes acting at different levels of the pharmaceutical nexus (Petryna and Kleinman 2006) and according to specific cultural, economic, and political factors. In other words, the politically, economically, and culturally charged social space in which products circulate affect and shape the circuits of the products, as well as their very nature (composition) and meaning. This is what Adams et al. (2010, 4) in their work on Tibetan medicine have come to name "Sowa Rigpa sensibility" by which they mean a medicine's capacity to be shaped and transformed, adapting to local needs and expectations, while still holding fast to a coherent set of principles that define its epistemological foundations. In this flexibility resides not only the "fragility of objecthood" (Appadurai 2006, 15), but also the potential for its resilience.

The flexibility of Asian medicines and their transformation engendered by their inscription in this complex system of governance often lead to adjustments in the regulations which in turn engender further transformations of the products, whether in their contents, their labels, or the actors involved in their circulation.

Now, if the flexibility of the products – and to a certain extent, the regulatory system – grants medicines an expanded circulation, it is undeniable that in many cases, the circulation is actually reduced. Indeed, as already shown by Craig (2012) and Saxer (2013), some regulations seem to bring attempts both to the availability of raw materials for manufacturers and to the accessibility of medical products to consumers. The intensive planting, harvesting, and collection of resources by big companies operating according to purely financial interests is leading to a scarcity of resources which severely affects small manufacturers and traditional healers who used to get their material from the wild (Jagtenberg and Evans 2003, 323). This in turn affects a large part of the population, mainly the poor and those living in remotes areas for whom small manufacturers and healers used to be the main – if not the only – available providers of medicines. If regulations such as border regimes, as well as licensing and taxing systems, do not touch at the material per se, at least not directly, they still impact on its accessibility through their action on the circulation of material and people involved in production, trade, and distribution. As noted by Saxer (2013, 231), by imposing taxation systems and efficacy and safety checks, border regimes restrict and channel the transit of people and goods across the border. Licensing systems also orient circuits by fostering certain distribution avenues and hindering others: traditional medicines are increasingly being distributed through pharmacies and big medical shops, which sell them directly to the customers, and less by small shops and healers. Finally, the combination of GMP, licensing systems, and border regimes further reduce the accessibility of medicines by making them more costly. This is particularly problematic for economically disadvantaged people living in remote areas for whom these healers provide the only access to healthcare. This suggests that if the different regulations at the basis of medical governance were originally aimed at granting the national and international population greater accessibility to safe and good quality medicines, this aim is actually barely reached.

The book: toward a multi-layered and cross-national approach

The relationship between circulation and governance is thus an intimate and complex one, one which is grounded on medicines-as-things. It is indeed the *thinginess* (Appadurai 1986) of medicines themselves – i.e., their mobility, fluidity, and capacity to cross borders, and also their medical, economic and political value – that allows for their circulability at the same time as it calls for their regulation. This regulation is produced and enacted by actors who are situated at different levels where "'level' here refers to the international, national, regional and local tiers of social organization (see Van der Geest 2011) – in which political power, commercial power, and individual agency play prominent roles" (Meier zu Biesen, Chapter 6 of this volume). Now this hierarchy is reflected in and reproduced by the agency the different actors have on the circulation and regulation of the medical products. This means that regulatory practices take different meanings and have different implications for different people, depending on the position they occupy within this assemblage: if some will benefit from it, others will suffer from it. As expressed by Lee and LiPuma through the notion of "culture of circulation" (2002), circulation is "a generative process that entails negotiation at the nodes as well as the performative channel making that enables flows, and hence creates spaces of new therapeutic opportunities and (new) forms of exclusion, control, and restrictions" (Meier zu Biesen, Chapter 6 of this volume).

Despite the relevance of this relationship for Asian medicines, it has seldom been examined in all its complexity in the literature. With the present book, we intend to contribute to the understanding of this relationship by examining the assemblage that constitutes it and the power structure that shapes it.

This book emerges from the multidisciplinary workshop "Governance and Circulation of Asian Medicines"[2] which took place in September 2015 at the Asia Research Institute of the National University of Singapore, which also convened the event. The book was crafted during a three-year dialogue between the different contributors, most of whom participated in the workshop, while others joined the project successively.

Very much like the workshop itself, the book mainly aims at unpacking the organized sets of practices that govern contemporary Asian medicine from their production in the lab to their circulation within circuits and networks of all kinds. We intend to examine the plurality of actors involved in such governance from local to global, public to private institutions, and the diverse, converging, or conflicting logics actors follow, the negotiations and tensions that might rise, and the consequences for products, people, and knowledge alike.

This book also aims to look at how these regulations, dominated by biomedical regimes, a new form of biopower, engender a transformation of the products, their circuits, and actors involved, thus revealing the fluid and multiple nature of mobile commodities (Appadurai 1986) in general and the adaptability and sensitivity (Craig 2012) of Asian medicines in particular.

Although they all examine the relationship between governance and circulation, the different contributors differ in the products they examine or, rather, in the categories of products they focus on, such as "traditional medicines" (Coderey, McNamara, Smith; Chapters 3, 1, and 4, respectively), "biomedical drugs" based on "traditional medicines" (Meier zu Biesen, Chapter 6), "health products" (Chee, Wang; Chapters 5 and 7, respectively), "health functional food" (Ma, Chapter 2). Those that look at products that are mainly of vegetal origin may include in their analysis other ingredients such as metals (Coderey, Chapter 3) and animal parts (Chee, Chapter 5). The authors also differ in their focus on the circulation of medicines within one country (Ma, Coderey, McNamara; Chapters 2, 3, and 1, respectively) or across countries (Meier zu Biesen, Wang, Chee, Smith; Chapters 6, 7, 5, and 4, respectively), as well as the nature of governance–legal (McNamara and Chee; Chapters 1 and 5, respectively), economical (Meier zu Biesen and Chee; Chapters 6 and 5, respectively), and political (Smith, Coderey and McNamara; Chapters 4, 3, and 1, respectively). They also vary in the scale of governance they observe – from a totalitarian form, covering the whole society through action targeted on individual bodies (Smith, Chapter 4), to a form mainly limited to the medical sphere (other chapters). Finally, they differ in the actors they focus on: some mainly look at the top-down aspect of governance, with the imposition of national and international standards as central sites of power (Petryna and Kleinman 2006; Kumar and Dua 2016) (Smith, Meier zu Biesen, Ma, Chee, McNamara, and Coderey; Chapters 4, 6, 2, 5, 1, and 3, respectively), while Wang (Chapter 7) focuses on response or alternative governance from below.

Each author illustrates the relations between different actors and objects, and their embedment with an assemblage of social forces animated by relations of complementarity and hierarchy. For instance, looking at practitioners, manufacturers and other local actors, Wang, Meier zu Biesen and myself (Chapters 7, 6, and 3, respectively) show how the choice to implement regulations or otherwise, to embrace them or to bypass them, and the way to do it is the outcome of a complex negotiation between medical, political, and social forces, an assemblage which aggregates a wide range of elements ranging from herbs, machines, money, architectural arrangements and packaging materials, knowledge, values, technology, money, laws, policies, (religious) ethics, design preferences, and certainly, social positions which determine the margins of choice and of negotiation available to one.

It should also be noted that no study on the circulation and governance exists which looks at different Asian countries and different Asian medicines, wherever they are practiced. The majority of studies on Asian medicines have focused either on a single country or on one kind of medicine or product. For the former, the interest was in the examination of how the local regulations have been shaped by global norms and biomedical standards (Adams 2001, 2002 on India; Hsu 2009 on China; Saxer 2013 on Tibet; Craig 2012 on Nepal; Wahlberg 2014 on Vietnam). For the latter, the interest was in transnational circuits of one specific form of medicine and its "worldling" (Zhan 2009), how it would take a different shape depending on the local contexts, and how these changes would then affect the medicine's changes within the country (Pordié 2008, 2014, 2015; Wujastik and Smith 2008 on India; Hsu 2002 on Tanzania; Zhan 2009 on China).

The cross-national and cross-medicines approach we have chosen to embrace aims at allowing for comparison and for highlighting different countries' political and economic histories at the same time as it opens up the possibility of a reflection on the region and its place in the globalized world – and thus on the specificity of the circulation and governance of Asian medicines. In this volume, we bring together studies of different Asian countries, with very different histories and levels of economic development (from Bangladesh and Myanmar to Korea and Singapore), and different levels of exposure to, or integration into, the global influence where traditional or herbal medicine has reached different levels of development and circulation (from mainly national in Myanmar to highly international with China). It is also one of the strengths of this volume that it includes countries such as Bangladesh, Myanmar, and Singapore that are greatly underrepresented in the literature on Asian medicines.

Thanks to this multiplicity of approaches, the different chapters bring a multilayered contribution to the study of governance and circulation of Asian medicines.

Main themes across the chapters

Biomedical governance and the power of science

Most of the chapters examining Asian medicines in the context of globalization stress that the regulatory system is dominated by biomedicine, and that it is biomedicine which defines the basic criteria for the evaluation of the quality, the safety, and the efficacy of medicines through guidelines in "good practices" and binding standards (Pordié and Gaudillière 2014). They also discuss how this biomedical governance poses both ethical and practical problems for Asian medicines, whose paradigms differ from biomedicine (Adams 2001, 2002; Craig 2012; Saxer 2013; Pordié 2008, 2010, 2011; Wahlberg 2008b, 2012).

We intend to enter in dialogue with this set of works and to push the reflection further. If many studies so far have stressed the fact that local authorities and manufacturers embrace biomedical standards and have discussed the consequences of this choice, less attention has been given to the *whys* and the *hows* of this choice, besides the fact that biomedicine has become globally dominant. This question appears to us all the more relevant, given that the actors involved do have agency and could, potentially, make a different choice.

In the case of alignment with biomedical standards, one should ask what these regulations mean for local governments. Previous works on China (Taylor 2005; Hsu 2009), Vietnam (Wahlberg 2014), and other countries have suggested that the biomedicalization of the medical arena was very much part of the project of modernization and development of the country which went along the nation-building. The same is true also for other countries covered in this book, notably Korea (Ma, Chapter 2), Myanmar (Coderey, Chapter 3), and Singapore (Smith, Chapter 4). The latter author also shows how in Singapore, the phenomenon took an economic twist which became very impactful on society as a whole. Indeed, governing medicine was part of the bigger project of governing the nation in the name of economic efficiency, and this was done by disciplining bodies and minds. In the author's words:

> Biomedical research and development was encouraged as an important economic growth sector – designed to attract international investors, bio-pharmaceutical companies, and biomedical professionals, and to promote Singapore's image as a "modern," high-tech nation-state. A productive body politic – surveilled, disciplined, and categorized – was engineered as the country's primary natural resource and embodiment of the "Biopolis of Asia."

This spirit, Smith continues, translated into traditional medicine through the shaping of medical spaces into "efficient healing environments that channeled flows of people, ideas, practices, and materials in accordance with state agendas seeking to maintain the economic and biological vitality of the population."

In tune with this, my own Chapter 3 suggests that in Myanmar, the ideals of science and modernity on which biomedical governance relies, as well as the standardization and secularization entailed by GMP, are in compliance with the states' agendas related to nation-building, the development of a modern nation, and the defense of national culture. More importantly, I show what this meant on the ground: that the regulation of medicine through its biomedicalization aimed not only to comply with biomedical standards accepted internationally, but also to neutralize aspects of medicine perceived as a potential threat to the authority of the central government – be they related to minority groups or to supernatural-magical aspects.

Equally interesting are cases of non-alignment, such as the case of the *tso-thal* previously described by Saxer (2013, 71–75), because they reveal how other forces and actors are at stake beside biomedicalization and the WHO. *Tsothal* is the main ingredient of the famous Tibetan "precious pills." It contains purified

mercury and gold, and its preparation includes complex rituals combining meditation and chanting. Despite its strong religious component, which is problematic
for the Chinese state, and despite its mercury and gold content, in opposition
to Chinese GMP and drug laws, Chinese regulations are not applied at all to its
preparation. This is because *tsothal* is recognized as part of Tibetan cultural and
scientific heritage and since 2006, a part of China's intangible cultural heritage.
Another major reason is that the precious pills market is a multi-billion-dollar
business in China. In this volume, I make a similar case for Myanmar, where the
fact that alchemy is recognized as part of the local medical heritage of the country
justifies/explains the fact that it is maintained in the official and standardized form
of traditional medicine, even though it contrasts with the biomedical criteria of
what medicine is. Moreover, in Myanmar as in China, the circulation of alchemic
products is also supported by the weakness of the control system. In Myanmar,
however, my chapter suggests, there is also a political reason behind the resilience
of alchemy, insofar as alchemy represents a challenge – if not a vengeance against
the West and its medicine.

Shifting attention from the state to the manufacturers, sellers, or border officers,
contributors look at how these actors respond to regulations and the logics they
follow, and try to identify the factors which favor or hinder the implementation of
the intended regulations. If compliance – or the absence of it – is usually mainly
driven by economic reasons, as we will discuss later in this introduction, it is also
usually grounded in a critical engagement with the biomedical paradigm. Manufacturers and sellers ponder whether the adaptation would favor or hinder the
efficacy and safety of the product, and whether it would change its identity. This
is all the more so, considering that the way they understand the products can differ from the WHO's perspective. Most agree that something is lost in the adaptation process – notably through the reduction of ingredients and the neutralization
of magico-religious aspects (Adams 2001, 2002; Adams et al. 2010). However,
many do not understand this process in any romantic terms, but consider it as a
central feature of contemporary transformations of Asian medicine. For instance,
Laurent Pordié and Jean-Paul Gaudillère (2014) have previously shown through
the example of the "reformulation regime"[3] how Ayurvedic medicine has been
able to turn the biomedical dominance into its own advantage; manufacturers
have reinvented their products by integrating new elements from the biomedical
paradigm, in order to make them more efficacious and more marketable.[4]

The biomedicalization of Asian medicine does not, however, operate only at
the level of the ingredients, but extends to the context and channels of the circulation of the products. Asian medicines are now often sold in pharmacies and shops
which also sell biomedical drugs, as shown by myself, Wang, and Smith (Chapters 3, 7, and 4, respectively). The fact that the privilege of selling biomedical
drugs is not granted to shops of traditional medicine reflects and reinforces once
again the hierarchization of the two medical systems.

Insofar as the literature has mainly focused on the need for traditional medicines to change in order to comply with biomedical standards, and being thus
granted a certain legitimization and interrogated to what extent a medicine can

change without compromising its efficacy and identity (Adams 2002; Saxer 2013), less attention has been given to the question of safety and the legitimacy of science in assessing traditional medicines' safety. In her Chapter 2 on Korea, Ma examines the relationship between safety and science, and reflects on whether scientific methods can be employed as a safeguard to ascertain the safety and efficacy of health products; if a regulatory agency serves as gatekeeper to keep adulterated foods from entering the consumer market; and, finally, if the regulatory body might employ sound science either to falsify or corroborate alleged scientific claims made by multiple parties with conflicts of interest.

The power of invisible ingredients

An aspect that is often overlooked in the literature, and that we want to bring back to scholarly attention, is the resilience of certain spiritual or religious components. If in some cases, these aspects are marginalized by regulations like those found in China, where every expression of religion is banned, in most cases they are marginalized by the mechanization and rationalization of the production process and the commodification and the commercialization of the products. Regardless of the specific reasons behind the marginalization, in many cases, these factors persist.

Such is the case of mantra recitation I illustrate in Chapter 3. Even though mantras are discriminated against because of their non-scientific nature and especially because of their magical or religious connotation, the fact that they can be recited discreetly and their outcome is "invisible" make their maintenance quite easy. Spiritual, religious, and other traditional symbols often also appear on the packaging or in advertisements that serve as a testimonial of commitment to tradition or as a strategy to facilitate circulation. Hence, if the galenic (or ready-to-use nature) and the shape of the medicine as much as the inscriptions on the packaging – with the list of ingredients and the name of the symptoms, which may be even in biomedical terms – reassure the customer that he/she is purchasing a proper modern – and thus safe and of quality – medicine, names, logos, and images satisfy the thirst for tradition and spirituality. It is the hybrid character of the medicine – its blending of what local people perceive as representing modernity and tradition – which grants its success (Van der Geest and Whyte 1989).

All those often-overlooked intangible aspects are particularly important for Asian medicines, given their common entanglement with religion and their deep identity and often nationalistic attributes that greatly appeal to the West.

The role of the economy in governance and circulation of Asian medicines

One of the main forces animating and shaping the relationship between circulation and governance is the economy, in tune with, and at the service of, the seemingly unbreakable intertwinement with a monolithic and homogenizing form of capitalism which globalization has – so far – brought about (Jagtenberg and Evans 2003, 323). At all levels – global, international, national, regional, and individual – whether in the private or in the public sector, attempts to govern the circulation of medicines, and choice of whether or not to comply with regulations, are greatly influenced by economic concerns and agendas. Such decisions then shape the circuits through which medicines circulate.

If we can assume that the WHO does not make decisions on the basis of its own economic interests, it is undeniable that this organization can easily become – and actually often is – an ally of other major actors – states and pharmaceutical companies *in primis* – which do act according to a specific economic agenda. The authority and weight of the WHO provides its allies a certain veneer of legitimacy, respectability, and neutrality, while the circulation which results from such alliances is often morally problematic insofar as it leads to forms of monopoly and exclusion. This phenomenon is clearly illustrated in this book in the work of Meier zu Biesen on artemisinin in Tanzania (Chapter 6). The author shows that the WHO has agreed to allow Novartis to be the only company able to supply malaria treatments to the WHO and the Global Fund. Novartis is indeed the only company which has a patent for this medicine. The monopoly of Novartis is strategically justified by the reliance on scientific standards embodied and granted by WHO. As stated by the author, "the economic interests of the pharmaceutical industry are 're-declared' through WHO validation processes, which are based upon the standards developed by these very same actors." This monopoly has enormous consequences at all levels. First of all, it restructures the social life of the products, stretching it across several countries. Industries in Tanzania produce artemisinin using recrystallization from the vaporized *Artemisia* raw extract – a viscous oil – and process it further into artemisinin crystals. The crystals are distributed to Novartis, combined with lumefantrin in China, India, and the United States, and find their way back into developing countries as the combination drug Coartem®. Second, the monopoly held by big pharmaceutical company reduces the accessibility of medicines by increasing the price of patented medicines and making others illegal and non-accessible. This relates to the fact that as shown by Petryna (2009), donors – not recipients – tend often to predominate, and the operations of international (health) organizations tend to reinforce existing and unequal power relations between countries. This situation is paradoxical, given that artemisinin is available in the country and there would be possibility for local companies to produce the medicine at low cost and thus make it easily accessible to the population. The author adds that the problem is not only in the high cost of the patented medicine, but in the fact that the price of artemisinin fluctuates, meaning that the salary processing factories can afford to pay farmers for dried leaves varies from time to time. She also highlights that these power relations are maintained also through the control of the circulation of knowledge relative to the products. In other words, the way such knowledge is channeled toward farmers and producers favors the reproduction of the hierarchies. This chapter thus shows how the standardization of medicines appears as a necessary condition for biomedical methods of production but on the other hand, it facilitates the development of monopolies over strategic knowledge.

Another example where the WHO is used as a symbol of scientificity and neutrality, and is thus manipulated by other actors for political and economic purposes, is given by McNamara in her Chapter 1 on Bangladesh. The author illustrates how "although WHO might think of itself as a politically neutral organization that gives technical support, its guidance on drug and health policies often became entangled in local political processes." Indeed, the representatives of medical traditions that had been marginalized by the state use the symbolic neutral power of the WHO to bolster support for their own practices and medicines. This need for political and legal recognition is also (mainly?) motivated by a desire for economic advantages.

At the level of the state, decisions are often economically driven, in the sense that they are intended to protect and boost the local economy. For instance, Ma shows in Chapter 2 that the Korean state decided to promote scientific research to turn herbal medicines into biomedical drugs because of the high success and circulability of these medicines. In a similar way, Chee shows in Chapter 5 how the circulation of fish liver oil was regulated on the basis of economic factors (the tariff law in the United States and the health insurance in the United Kingdom), and not on scientific criteria or because of medical concerns.

The state does not, however, always possess much agency in the regulation and circulation of medicines. Indeed, in the case of Tanzania as discussed previously, for instance, the state is actually powerless; its power is silenced, paralyzed by the regulatory regime imposed by the alliance between Novartis and WHO.

The factors determining the actions of traders, manufacturers, practitioners, and vendors are also economic in nature. The agency these actors have on the circulation of medicines is not unlike the freedom they have to choose whether or not to comply with regulations, and the choice they actually make often greatly depends on their economic power. Indeed, as I show in Chapter 3 on Myanmar, not everyone has the financial resources to purchase equipment and ingredients necessary for the production of medicines in compliance with GMP, let alone for the payment of forms, controls, licenses, and taxes. Both my Chapter 3 and Wang's Chapter 7 show that if the choice to comply with regulations is often motivated by the desire to expand one's circuits, the same can be true for choosing to bypass the regulations. I also illustrate that the decision of non-compliance is for some political, and for others motivated by therapeutic concern and the desire to keep producing affordable medicines (usually referred to as "moral economy"). Moreover, my chapter highlights that more often than not, the situation is not black or white, but somewhere on a wide spectrum between compliance and non-compliance – and that this often has to do with the agency left to local actors by the fluidity, flexibility, and even blurredness of the governance system (Saxer 2013; Quet et al. 2018). Many manufacturers opt for a midway solution of compliance for some products and non-compliance for others, as already shown by Saxer in China and reiterated here with the case of Myanmar. In both cases it would seem that compliance usually grants access to greater (licit) circuits and improved financial gains, while non-compliance is often motivated by the wish to protect the efficacy and identity of one's medicine, even though this is often done because certain profits result from it. In any case, licit circuits often come to cover illicit ones.

Economy also explains other transgressions within the regulatory system. Most bypasses – omissions of safety and quality checks, allowance of illegal transborder trade, etc. – happen because of financial compensation – bribes – or because of special connections with the authorities, often translated into a gift exchange between goods, on the one hand, and benefits and protection, on the other hand (Coderey, Chapter 3).

Governing categories

If the circulation and regulation of Asian medicines is primarily related to the materiality of the objects – in terms of ingredients and in terms of economic value – it is also greatly related to something quite intangible: their names, definitions, and the categories they belong to: "traditional medicines," "biomedical drugs" produced from "traditional medicines," "herbal medicines," "health functional foods," or "health products" and even "food." The way medical products are regulated and circulated depends, first of all, on their categorization. Categorization is used to make order, to qualify and disqualify by creating boundaries, and by grounding the legitimacy of such acts on the "neutral" authority of "science" embraced by the major global and state actors. Categories have the power to open or close circuits. In other words, labeling and categorizing are, once again, an act of power – a tactic to facilitate or hinder a circulation.

Distinguishing between traditional medicine and biomedical drugs is at the core of the regulation process. Indeed, it is a fact that in principle, traditional medicine differs from biomedicine which represents the reference for quality and safety, and this creates the necessity for an adaptation of traditional medicine. Depending on how far the adaptation goes, the limits between the two forms of medicines can become blurred. This is what happens, for instance, with Ayurvedic medicines through the use of the "reformulation regimes" (Pordié and Gaudillière 2014). Something analogous is described by Ma in Chapter 2 for Korea, where herbal medicines are turned into biomedical drugs through the principle of "reversal pharmacology." Interestingly, Ma adds that the blurring of the boundaries favors some and upsets others, igniting "deep-rooted conflicts between the supporters of two medical/pharmaceutical systems, i.e., Korean medicine and Western biomedicine." Similar is the example described by Meier zu Biesen in Chapter 6 of the Chinese plant *Artemisia* used to produce the anti-malaria drug Coartem®.

The power of categorization is also shown by McNamara, who describes in Chapter 1 how traditional medicine in Bangladesh has been greatly affected by the recommendations to incorporate traditional medicine into primary healthcare and to provide access to "essential drugs." The introduction of the notion of essential drugs inspired new drug policies that led to disagreements over how traditional medicine was categorized and practiced. Indeed, because of this list, too, traditional medicine has been regulated according to biomedical principles.

The other main categorization is between medicine and food. The boundary between food and medicines has always been blurred as reflected in the saying common across a number of Asian societies that "food is medicine and medicine is food." The regulations applied on medicines are much more complex and strict compared to those applied on food, and in many countries, to be registered as medicine, a product needs to be tested through scientific methods that apply also to biomedical drugs. The fact that products which are quite indistinguishable are subject to different treatment has led to the emergence of the ambiguous category "health products." As stated by Chee in Chapter 5, this term refers to

> those non-prescription pharmaceutical-like substances that are advertised as supporting general well-being, and thus do not require in most instances a doctor's prescription. Some "health products" originated as drugs before being re-categorized by states, either through regulation or because their producers (or users) were responding to regulations which imposed a financial cost. In other cases, they were more food-like to begin with and took on medical characteristics later.

The resulting rules or policies have had an effect on how the products have circulated, and how the products have been influenced in turn by patterns of circulation.

Chee and Wang examine how, in order to facilitate the circulation of some products, states and manufacturers choose to categorize these products as a "health product" or "food supplement," something they can do because of the liminality of these products, being between drug and food. Comparing the cases of China, the United Kingdom, and the United States, Chee shows in Chapter 5 how taking advantage of the fact that fish liver oil is a "boundary object," the three states make use of the ambivalent category of "health product" in order to facilitate the circulation of the product. In the three governance regimes, the definition of the product is based not so much on arguments about intrinsic qualities or even use, but fiscal and political priorities. In the same vein, Wang shows in Chapter 7 that to be categorized and circulated as a drug in order to penetrate the French market, the claimed medical properties of a product must be proven through scientific, biomedical-based methods, and hence producers of Chinese medicines choose the category of health product. In this way, the product is sold in shops and without prescription, but dressed in specific packaging which reflects its hybrid identity.

Conclusion

If circulation means flowing across boundaries, governance is about creating, recreating, negotiating, contesting, and shifting boundaries, be them between countries, medicines, or people. At the crossroads between governance and circulation are multiplication, negotiation, and tension processes which involve a network of actors, objects, interests, and values, a balancing between gains and losses, a marriage or an arm wrestling between science and culture, modernity and heritage, the material and the symbolic. The outcome of this negotiation has several consequences on the efficacy, the quality and the accessibility of the medicine, and also on its meaning and symbolic power. This shows once again that medicines are much more than just a normal commodity and much more than just medical pills; but rather, they embody cultural, social, and political aspects as well. It also suggests that the system of governance is not only a source of movement and possibilities, but also a locus and instrument of exclusion, control, and restriction.

Sketching the individual chapters

Karen M. McNamara's Chapter 1 examines the regulations of traditional medicine in Bangladesh and discusses the role of the WHO in such regulations. Indeed, the development of national drug policies for traditional medicines in Bangladesh coincided with global priorities around the increasing role of pharmaceuticals in healthcare as promoted by institutions like the World Health Organization (WHO) and the World Trade Organization (WTO). The chapter shows how the WHO's recommendations to incorporate traditional medicine into primary healthcare systems and to provide access to "essential drugs" became entangled with contestations over how to categorize, produce, and circulate traditional medicine in Bangladesh. Traditional drug-manufacturing companies and the state both call on the authority of the WHO as a legitimate source of knowledge. McNamara shows that WHO has a contradictory role in institutionalizing traditional medicine in Bangladesh, revealed in its dual role as an actor that influences policies and as a symbol of a powerful global institution.

Eunjeong Ma's Chapter 2 examines in detail the public outbreak which took place in South Korea in April 2015 following the announcement made by the Korea Consumer Agency (KCA) that about 90% of functional foods containing *Cynanchum wilfordii* (*Baekshuoh*) on the market were not made out of "authentic" plants, thus raising panic about safety of the products and questions regarding regulatory culture and practices in the realm of health functional foods. Taking its point of departure from this case study, the chapter interrogates how the safety of natural health products is perceived, practiced, and managed in contemporary South Korea, and examines the use of scientific methods as a safeguard to ascertain safety and efficacy of health products and the role of a regulatory agency as gatekeeper to keep adulterated foods from entering the consumer market. This chapter unravels the complex, interconnected web of market, regulation, technologies, venture businesses, government, and consumers in South Korea.

Céline Coderey's Chapter 3 is a comprehensive ethnography of the system of regulation of traditional medicine in Myanmar, its diversified meanings and actual implementations. It examines how global forms are reterritorialized in this specific country by examining how they are understood, apprehended, and implemented by local actors – notably the state, manufacturers, and healers. The author shows that the regulations implemented in the country are the outcome of a re-appropriation of biomedical norms by the local government which adds to them new meanings and functions – notably the neutralization of individual, ethnic, familiar diversity, and the disenchantment of the practice in the name of building a modern Buddhist nation. It also shows that manufacturers and healers, far from blindly complying with these regulations, are critically engaging with them to ponder pros and cons, and eventually respond according to the specific position they occupy within the medical and social space.

Arielle A. Smith's Chapter 4 discusses the place and evolution of Chinese medicine in Singapore against the background of the country's 20th-century social and economic development, and the increased legitimacy given to the biomedical framework, and hence its domination. It shows that labeled "complementary" at best and "quackery" at worst, Chinese medical practices and materials circulate in somewhat strained relations to Singapore's biopolitical processes. This chapter describes various convergences and divergences of biomedicine and Chinese medicine within this sociopolitical milieu, and explores the contemporary modulation of the latter in accordance with dynamic sets of interests. It illustrates how fluidity, permeability, and complex power relations characterize both postcolonial Singapore and the practice of Chinese medicine therein.

Liz P.Y. Chee's Chapter 5 examines the question of the boundaries between the categories of food vs. medicine and the way the categorization affects the circulation. Using the case of fish liver oil and its regulation in three different countries – the United States, the United Kingdom and China – Chee shows that "health products" have emerged as a modern, yet ambiguous, category influenced directly or indirectly by state attempts to separate and regulate the categories "food" and "drug." The chapter suggests that it is political economy and not laboratory or science-based classification that determines how these borderline substances are categorized and re-categorized. It also shows how the ambivalence of the category facilitates the circulation of the product, making it an ideal marketing category.

Caroline Meier zu Biesen's Chapter 6 explores the interplay of circulation and governance of the plant-based pharmaceutical anti-malaria drug artemisinin. First, it unpacks the complex journey from the detection of artemisinin in China in the 1970s to commercial *Artemisia* plantations in East Africa and today's global annual administration of 400 million treatment courses of artemisinin-based combination therapies (ACTs). Secondly, using Tanzania as an example, it examines the pharmaceutical logic by which ACTs circulate and investigates how global health actors (WHO, national drug regulators, donors), commercial actors (pharmaceutical industry, private enterprises), scientists, and philanthropic institutions have created, controlled, and stabilized their pharmaceutical legitimacy. By investigating the pharmaceutical nexus driving the Artemisinin Enterprise, this chapter achieves a multi-layered look at the alliances between these actors, the regimes of intervention, and capital accumulation. This chapter touches important issues such as intellectual property, patents, and monopolies of multinational companies, the tension between interest/profit and concern for people's health, the hierarchy of knowledge, and the injustices toward farmers.

Simeng Wang's Chapter 7 examines the relations between transnational circulation and governance through the example of Chinese medicine circulating between China and France. It interrogates, on the one hand, how the circulation of Chinese medicine has been rendered possible between these two countries in the last decades' context of regulation, and on the other hand, how this circulation shapes the regulatory framework. The main focus is on practices and strategies adopted by product sellers and illegal practitioners to circumvent regulatory rules and to legitimize themselves professionally, including re-labeling of products, crossing of national borders, and professional networking. Through a three-step analysis (the sales of Chinese medicinal products, the professional legitimation of illegal practitioners, and the training conditions for future Chinese medicine practitioners), the findings illustrate how circulation and regulation affect one another, and demonstrate tight and diversified links between the ways the actors circumvent governance and how they legitimize themselves. This work beautifully shows how cross-country variation of regulation frameworks produces highly differentiated itineraries of Chinese medicine circulation.

Notes

1 Although often initiated by the states, the integration of traditional medicines within the healthcare systems, was soon supported by the World Health Organization (WHO), notably during the Alma-Ata conference of 1978, which saw there a chance to bridge Asian medicine and biomedicine, and to extend health coverage in both demographic and financial terms. In countries like India (Pordié 2008), Vietnam (Wahlberg 2014) and Myanmar (Coderey forthcoming 2019), this integration was part of the post-colonial nation-building process while in others, like in China (Saxer 2013; Craig 2012; Hsu 2009), it was part of a nation development-modernization program. In either case, the process was an initiative of local authorities characterized by strong nationalistic ethos, a desire to preserve the national culture and identity vis-à-vis the growing dominance of the West.
2 The event was co-organized by the Asia Research Institute, National University of Singapore, and the CERMES3 in Paris (in the framework of the ERC Project Globhealth).
3 This innovation reformulates complex polyherbal drugs which use medicinal plants as opposed to isolating chemical active ingredients. The major agents of this reformulation regime in India are large Ayurvedic drug companies and, increasingly, botanists and botanical gardens (Gaudillière 2014).
4 For more examples of this kind, see Pordié and Hardon (2015).

References

Adams, Vincanne. 2001. "The Sacred in the Scientific: Ambiguous Practices of Science in Tibetan Medicine." *Cultural Anthropology* 16 (4): 542–575.
Adams, Vincanne. 2002. "Establishing Proof: Translating Science and the State in Tibetan Medicine." In *New Horizons in Medical Anthropology: Essays in Honour of Charles Leslie*, edited by Mark Nichterand Margaret Lock, 200–222. London: Bergin and Garvey.
Adams, Vincanne, Mona Schrempf, and Sienna Craig, eds. 2010. *Medicine, between Science and Religion*. New York: Berghahn Publishers.
Appadurai Arjun, ed. 1986. *The Social Life of Things: Commodities in Cultural Perspective*. New York: Cambridge University Press.
Appadurai, Arjun. 1996. "Disjuncture and Difference in the Global Cultural Economy." Chapter 2 in *Modernity at Large: Cultural Dimensions of Globalization*. Minneapolis, MN: University of Minnesota Press.
Appadurai, Arjun. 2006. "The Thing Itself." *Public Culture* 18 (1): 5–21.

Banerjee, Madhulika. 2009. *Power, Knowledge, Medicine: Ayurvedic Pharmaceuticals at Home and in the World*. New Delhi: Orient Blackswan.

Bode, Maarten. 2002. "Indian Indigenous Pharmaceuticals: Tradition, Modernity, and Nature." Chapter 11 in *Plural Medicine, Tradition and Modernity, 1800–2000*, edited by Waltraud Ernst. London: Routledge.

Bode, Maarten. 2008. *Taking Traditional Knowledge to the Market: The Modern Image of the Ayurvedic and Unani Industry, 1980–2000*. Hyderabad, India: Orient Longman.

Burawoy, Michael, Joseph A. Blum, Sheba George, Zsuzsa Gille, Teresa Gowan, Lynne Haney, Maren Klawiter, Steven H. Lopez, Sean O'Riain, and Millie Thayer. 2000. *Global Ethnography: Forces, Connections, and Imaginations in a Postmodern World*. Berkeley: University of California Press.

Coderey, Céline. 2019 (forthcoming). "Myanmar Traditional Medicine: The Making of a National Heritage." Modern Asian Studies.

Collier, Steven J., and Aihwa Ong. 2005. "Global Assemblages, Anthropological Problems." In *Global Assemblages: Technology, Politics, and Ethics as Anthropological Problems*, edited by Steven J. Collier and Aihwa Ong, 3–21. Malden, MA: Blackwell.

Craig, Sienna. 2012. *Healing Elements: Efficacy and the Social Ecologies of Tibetan Medicine*. Berkeley: University of California Press.

Craig, Sienna, and Vincanne Adams. 2009. "Efficacy, Morality, and the Problem of Evidence in Tibetan Medical Research." *Complementary Therapies in Medicine*. doi: 10.1016/j.ctim.2008.03.003.2009.

Engelhardt, Ute. 2001. "Dietetics in Tang China and the First Extant Works of *Materia Dietetica*." In *Innovation in Chinese Medicine*, edited by Elizabeth Hsu, 167–191. Cambridge: Cambridge University Press.

Etkin, Nina L., and Paul J. Ross. 1982. "Food as Medicine and Medicine as Food: An Adaptive Framework for the Interpretation of Plant Utilization among the Hausa of Northern Nigeria." *Social Science & Medicine* 16 (17): 1559–1573. doi: 10.1016/0277-9536(82)90167-8.

Foucault, Michel. 1979. *Discipline and Punish: The Birth of the Prison*. New York: Vintage Books.

Foucault, Michel. 1988. "Technologies of the Self: Lectures at University of Vermont Oct. 1982." In *Technologies of the Self*, 16–49. Amherst: University of Massachusetts Press.

Foucault, Michel. 1991. "Governmentality." In *The Foucault Effect: Studies in Governmentality (with Two Lectures by and an Interview with Michel Foucault)*, edited by Graham Burchell, Collin Gordon, and Peter Miller, 87–104. Chicago: University of Chicago Press.

Gaudillière, Jean-Paul. 2014. "Herbalised Ayurveda? Reformulation, Plant Management and the 'Pharmaceticalisation' of Indian 'Traditional' Medicine." *Asian Medicine* 9: 171–205.

Hsu, Elizabeth. 2002. "The Medicine from China Has Rapid Effects: Patients of Traditional Chinese Medicine in Tanzania." *Anthropology and Medicine* 9 (3): 291–314.

Hsu, Elizabeth. 2009. "Chinese Propriety Medicines: An 'Alternative Modernity?' The Case of the Anti-Malarial Substance Artemisinin in East Africa." *Medical Anthropology* 28 (2): 111–140.

Jagtenberg, Tom, and Sue Evans. 2003. "Global Herbal Medicine: A Critique." *The Journal of Alternative and Complementary Medicine* 9 (2): 321–329.

Jane, Craig R. 2002. "Buddhism, Science, and Market: The Globalisation of Tibetan Medicine." *Anthropology & Medicine* 9 (3): 267–289.

Kuhn, Thomas. 1970. *The Structure of Scientific Revolutions*. Chicago: University of Chicago Press.

Kuhn, Thomas. 1982. "Commensurability, Comparability, Communicability." *PSA: Proceedings of the Biennial Meeting of the Philosophy of Science Association*: 669–688.

Kumar, Nandini K., and Pradeep Kumar Dua. 2016. "Status of Regulation on Traditional Medicine Formulations and Natural Products: Whither Is India?" *Current Science* 111 (2): 293–301.

Latour, Bruno. 1988. *The Pasteurization of France*. Cambridge, MA: Harvard University Press.

Latour, Bruno. 1993. *We Have Never Been Modern*. Translated by Catherine Porter. Cambridge, MA: Harvard University Press.

Latour, Bruno. 2000. "When Things Strike Back: A Possible Contribution of 'Science Studies' to the Social Sciences." *British Journal of Sociology* 51 (1): 107–123.

Law, John. 2004. "And If the Global Were Small and Non-Coherent? Method, Complexity and the Baroque." *Society and Space* 22: 13–26.

Lee, Benjamin, and Edward LiPuma. 2002. "Cultures of Circulation: The Imaginations of Modernity." *Public Culture* 14 (1): 191–213.

Ma, Eunjeong. 2010. "The Medicine Cabinet: Korean Medicine under Dispute." *East Asian Science, Technology and Society* 4: 367–382.

Nguyen, Phuong Ngoc. 2012. "The Post-*Đổi Mới* Construction of Vietnamese Materia Medica: The Case of the Pharmaceutical Company Traphaco." In *Southern Medicine for Southern People: Vietnamese Medicine in the Making*, edited by Laurence Monnais, C. Michele Thompson, and Ayo Wahlberg, 179–202. Newcastle upon Tyne: Cambridge Scholars Publishing.

Petryna, Adriana. 2009. *When Experiments Travel: Clinical Trials and the Global Search for Human Subjects*. Princeton: Princeton University Press.

Petryna, Adriana, and Arthur Kleinman. 2006. "The Pharmaceutical Nexus." In *Global Pharmaceuticals: Ethics, Markets, Practices*, edited by Adriana Petryna, Andrew Lakoff, and Arthur Kleinman, 1–32. Durham: Duke University Press.

Pordié, Laurent. 2008. "Tibetan Medicine Today: Neo-Traditionalism as an Analytical Lens and a Political Tool." In *Tibetan Medicine in the Contemporary World: Global Politics of Medical Knowledge and Practice*, edited by Laurent Pordié, 3–32. London and New York: Routledge.

Pordié, Laurent. 2010. "The Politics of Therapeutic Evaluation in Asian Medicine." *Economic and Political Weekly* 45 (18): 57–64.

Pordié, Laurent. 2011. "Savoirs thérapeutiques asiatiques et globalisation." *Revue d'Anthropologie des Connaissances* 5 (1): 1–12.

Pordié, Laurent. 2013. "Spaces of Connectivity, Shifting Temporality: Enquiries in Transnational Health." *European Journal of Transnational Studies* 5 (1): 6–26.

Pordié, Laurent. 2014. "Pervious Drugs: Making the Pharmaceutical Object in Techno-Ayurveda." *Asian Medicine* 9 (1–2): 49–76.

Pordié, Laurent. 2015. "Hangover Free! The Social and Material Trajectories of PartySmart." *Anthropology & Medicine* 22 (1): 34–48.

Pordié, Laurent. 2016. "The Vagaries of Therapeutic Globalization: Fame, Money and Social Relations in Tibetan Medicine." *International Journal of Social Science Studies* 4 (2): 38–52.

Pordié, Laurent, and Jean-Paul Gaudillière. 2014. "The Reformulation Regime in Drug Discovery: Revisiting Polyherbals and Property Rights in the Ayurvedic Industry." *East Asian Science, Technology and Society: An International Journal* 8: 57–79.

Pordié, Laurent, and Anita Hardon. 2015. "Drugs' Stories and Itineraries: On the Making of Asian Industrial Medicines." *Anthropology & Medicine* 22 (1): 1–6.

Quet, Mathieu, Laurent Pordié, Audrey Bochaton, Supang Chantavanich, Niyada Kiatying-Angsulee, Marie Lamy, and Premjai Vungsiriphisal. 2018. "Regulation Multiple: Pharmaceutical Trajectories and Modes of Control in the ASEAN." *Science, Technology & Society* 23 (3): 1–19.

Rabinow, Paul. 2003. *Anthropos Today: Reflections on Modern Equipment*. Princeton: Princeton University Press.

Saxer, Martin. 2013. *Manufacturing Tibetan Medicine: The Creation of an Industry and the Moral Economy of Tibetanness*. New York and Oxford: Berghahn.

Serres, Michel, and Bruno Latour. 1995. *Conversations on Science, Culture, and Time*. Translated by Roxanne Lapidus. Ann Arbor: University of Michigan Press.

Taylor, Kim. 2005. *Chinese Medicine in Early Communist China, 1945–63: A Medicine of Revolution*. London and New York: Routledge Curzon.

Temkin, Owsei. 2002. *On Second Thought and Other Essays in the History of Medicine and Science*. Baltimore: Johns Hopkins University Press.

Tsing, Anna Lowenhaupt. 2004. *Friction: An Ethnography of Global Connection*. Princeton: Princeton University Press.

Van der Geest, Sjaak. 2011. "The Urgency of Pharmaceutical Anthropology: A Multilevel Perspective." *Curare* 34: 9–15.

Van der Geest, Sjaak, and Susan Reynolds Whyte. 1989. "The Charm of Medicines: Metaphors and Metonyms." *Medical Anthropology Quarterly* New Series 3 (4): 345–367.

Wahlberg, Ayo. 2008a. "Above and Beyond Superstition: Western Herbal Medicine and the Decriminalizing of Placebo." *History of the Human Sciences* 21 (1): 77–101.

Wahlberg, Ayo. 2008b. "Pathways to Plausibility: When Herbs Become Pills." *Biosocieties* 3 (1): 37–56.

Wahlberg, Ayo. 2012. "Family Secrets and the Industrialisation of Herbal Medicine in Postcolonial Vietnam." In *Southern Medicine for Southern People: Vietnamese Medicine in the Making*, edited by Laurence Monnais, C. Michele Thompson, and Ayo Wahlberg, 153–178. Newcastle upon Tyne: Cambridge Scholars Publishing.

Wahlberg, Ayo. 2014. "Herbs, Laboratories, and Revolution: On the Making of a National Medicine in Vietnam." *East Asian Science, Technology and Society* 8 (1): 43–56.

Wahlberg, Ayo. 2018. *Good Quality: The Routinization of Sperm Banking in China*. Berkeley: University of California Press.

Whyte, Susan Reynolds, Sjaak Van der Geest, and Anita Hardon. 2002. *Social Lives of Medicines*. Cambridge: Cambridge University Press.

World Health Organization (WHO). 2018. "WHO in Southeast Asia." www.searo.who.int/about/mission/en/.

Wujastik, Dagmar, and Frederick M. Smith, eds. 2008. *Modern and Global Ayurveda: Pluralism and Paradigms*. New York: SUNY Press.

Zhan, Mei. 2009. *Other-Worldly: Making Chinese Medicine through Transnational Frames*. Durham: Duke University Press.

1 WHOse guidelines matter?

The politics of regulating traditional medicine in Bangladesh

Karen M. McNamara

While visiting the only government degree college for Unani and Ayurvedic medicine in Bangladesh, I noticed two dilapidated vehicles permanently parked in front of the attached hospital clinic.[1] The light blue WHO (World Health Organization) logos painted on the car doors were barely visible under the years of accumulated dirt shrouding the white vehicles. When I asked students and teachers at the school about WHO's involvement in the college, they mentioned support and funding, but gave no specific details about what kind of support and funding. The institutional memory seemed limited, probably due to the cyclical rotations of new batches of students and also new principals, a political appointment that changed with every government administration.

Searching for a more detailed version of the story of WHO and traditional medicine in Bangladesh, I met with Asif,[2] a staff member of WHO in Bangladesh for over nine years. He said that in matters concerning traditional medicine,[3] WHO worked mainly with the Drug Administration and the Directorate of Homeopathy and Traditional Medicine. Asif clearly explained that WHO does not establish institutions, but does provide support and some equipment. If WHO did not establish institutions, then why were there cars with the WHO logos abandoned in front of the hospital? According to Asif, from 1993–1998, there were 35 projects executed by WHO under the first Population and Health Project in Bangladesh. It was during this period that many vehicles were procured. Non-governmental organizations (NGOs) carried out much of the work for these projects, which mainly focused on primary healthcare, but expanded to include traditional medicine.

The money to fund WHO projects was not necessarily WHO money, but usually was from foreign development projects. Although marked with the emblem of WHO, the vehicles were not bought or owned by WHO. According to the Bangladesh Board of Unani and Ayurvedic Systems of Medicine, WHO provided a mere 0.38 million taka[4] for the promotion of traditional and alternative medicine in Bangladesh (Alam 2007, 243). Jaffar, a representative from the Government Ministry of Traditional Medicine, told me that the idea for the development of the degree college for Unani and Ayurvedic medicine[5] came from the idea of "Health for All" promoted at the 1978 WHO conference in Alma-Ata, but that the funding for the college came partially from Reimbursable Project Aid (RPA) from the World Bank. Like Asif, Jaffar emphasized that WHO money and support is only for "technical support." Around the same time, WHO also gave technical advice through a project on Bangladesh's Essential Drugs and Medicine for Health Project.

The WHO logos on the cars are a trace of WHO's influence in institutionalizing traditional medical practices in Bangladesh. This chapter draws on two years (2007–2009) of multi-sited ethnographic research on the herbal and traditional medicine pharmaceutical industry in Bangladesh. The time spent at colleges and schools of traditional medicine in Dhaka was interspersed with interviews and participant observation at local clinics, traditional drug-manufacturing companies, and local community health organizations. In addition, I interviewed government and WHO officials involved in creating drug and health policies in Bangladesh. This chapter examines WHO's role in the governance and circulation of traditional medicines in Bangladesh, especially as these medicines interacted with global institutions and local politics. The creation of national drug policies for traditional medicines in Bangladesh during the 1980s coincided with global priorities around the increasing role of pharmaceuticals in healthcare. These developments reflect a broader shift in international public health at that time, from work in prevention and clinical care to providing access to medications. The idea of individual nations working together on public health with other national governments was reconceptualized in the 1990s, when global organizations began to have a more powerful role than individual nations. Therefore, international health agendas transformed into global (supranational) health issues that organizations like the WHO could take part in by promoting policies about health needs on regional and global scales (Brown et al. 2006). In this context, public health became more decentralized and pharmaceuticalized, creating a new form of "pharmaceutical governance" (Biehl 2006). The agenda of WHO also shifted as it interacted with global market forces, global health and development agendas, and local health policies.

These new forms of global governance and the influence of a neoliberal economic system also changed the terms of how medicine is categorized, produced, and circulated in Bangladesh. Government policies became less concerned with the development of institutions of traditional medicine and more concerned with the industrial production of traditional medicines. This commodification and production of medicines was not new. The large-scale manufacture of traditional medicine began in the Indian subcontinent during the late 19th and early 20th centuries (Habib and Raina 2005; Banerjee 2009; Bode 2002, 2008), and there has been a widespread growth in Asian industrial medicines across all regions of Asia (Pordié and Hardon 2015; Afdhal and Welsch 1988; Craig and Glover 2009; Wahlberg 2008; Gaudillière 2014). As Gaudillière (2014) points out, the process of pharmaceuticalization is not solely a biomedical process of mass-producing chemical drugs based on a molecular paradigm. He argues that pharmaceuticalization also happens to traditional medicines, specifically focusing on drug-innovation practices of Ayurvedic medicines in India. This innovation reformulates complex polyherbal drugs which use medicinal plants, as opposed to isolating chemical active ingredients. Large Ayurvedic drug companies and, increasingly, botanists and botanical gardens are the major agents of this reformulation regime in India.

The medical landscape of Bangladesh is similar to India because of a shared cultural and colonial history. The legitimacy of traditional medicine is tied to the plurality of healing systems, where colonial and state institutions have given more support to biomedicine (known as allopathy in Bangladesh) than to traditional medical practices. Bangladesh is a much smaller country than India and doesn't have as many Ayurvedic companies because many Hindu Ayurvedic practitioners and manufacturers fled to India from East Pakistan after the Partition of 1947.[6] The growing nationalist movements in colonial India tried to revive traditional medicine along religious lines, to associate Ayurveda with Hinduism and Unani with Islam. Today, the religious identities of Unani and Ayurveda in South Asia are fluid and contested, and many consumers, practitioners, and manufacturers do not subscribe to these rigid religious associations (Hardiman 2009, 281). However, others specifically appeal to religious sentiments to market their medicines. Also, since the 1980s, the number of Unani companies in Bangladesh has increased, coinciding with the growth of Islamic nationalism and the liberalization of the economy. In 2009, there were officially 261 Unani, 161 Ayurvedic, and 246 allopathic companies registered with the government as production units. Some companies produce only Unani medicines, others Ayurvedic, and still others produce both. These companies range in size, scale of production, targeted markets, and influence. A government worker did warn me that the numbers are somewhat misleading because some registered companies exist in name only or do not actually have a production unit. In my conversations with traditional drug manufacturers, their concerns focused more on the production and sales of medicines, rather than reformulating medicines. Perhaps this is because sales are targeted mostly at the domestic market or unregulated foreign markets, and few companies are equipped to focus on research and development.

WHO guidelines play a role in the governance and circulation of traditional medicines in Bangladesh, but they are not the only ones that matter because other actors and forces are involved in these processes. This chapter illustrates how boundaries around different types of medicine in post-colonial Bangladesh are influenced by WHO public health recommendations, as well as by World Trade Organization (WTO) policies on drug patents. I attempt to untangle this complicated story of traditional medicine, as it interacted with these organizations and market forces in the local cultural and political landscape. Although WHO might think of itself as a politically neutral organization that gives technical support, its guidance on drug and health policies can become entangled in local political processes. WHO's influence on traditional medicine in Bangladesh was not always direct, but its recommendations to incorporate traditional medicine into primary healthcare systems and to provide access to "essential drugs" inspired new national drug policies that led to disagreements over how traditional medicine was categorized and practiced.

Classification systems are produced through historically and politically contingent processes. What medical categories symbolize and what material effect they have on lives is malleable. In Bowker and Star's (1999) classic work, *Sorting Things Out*, they emphasize the importance of studying classification because:

> many scholars have seen categories as coming from an abstract sense of "mind," little anchored in the exigencies of work or politics. The work of attaching things to categories, and the ways in which those categories are ordered into systems, is often overlooked.
>
> (Bowker and Star 1999, 286)

I argue that WHO has a contradictory role in institutionalizing traditional medicine in Bangladesh, revealed in its dual roles as an actor that influences policies and as a symbol of a powerful – but "neutral" – global institution. The stories of the symbolic power of the WHO gets pushed to the shadows, much like the WHO logos hidden behind the dirt of the cars. Although official WHO categories set limits as to what counts as traditional medicine, I show how people representing these traditions in Bangladesh used this symbolic categorical acknowledgment from the WHO in order to bolster support for their own practices and medicines that had been marginalized by the state. Representatives of traditional medicine used this symbolic power of the WHO to rally support for their medical practices in the face of weak state support. My analysis centers around two main events to illustrate these complex processes. One is the creation of a National Drug Act in 1982 that regulated traditional medicine for the first time in Bangladesh, and the second is related to the more recent controversies around WTO patent rules and the categorization of herbal medicines vis-à-vis traditional medicines in the last 15 years.

WHO and "Health for All"

In 1975, the WHO and the United Nations Children's Fund (UNICEF) jointly published a report called *Alternative Approaches to Meeting Basic Health Needs in Developing Countries*. This report critiqued the vertical approach to healthcare which focused on specific diseases and the expansion of Western biomedical systems in developing countries. Instead, it called for examining the causes of morbidity, which were considered linked to poverty and ignorance. This model for public health was partially inspired by primary healthcare experiments in countries such as China and Bangladesh. In China, groups of Maoist "barefoot" doctors were deployed to provide healthcare in rural areas, using kits of both Western and traditional Chinese medicine (Greene 2011). In Bangladesh, the *Gonoshasthaya Kendra* (People's Health Centre) was set up by Zafrullah Chowdhury in 1972 to help the wounded during the war of independence with Pakistan. After the war, it continued to provide health and other services and its paramedics were the first to be trained outside of China. *Gonoshasthaya Kendra* "emphasized independent, self-reliant, and people-oriented developments" (Chowdhury 1995b, xi). Chowdhury eventually founded Gonoshasthaya Pharmaceuticals in 1981, and was a key advisor in the formulation of the 1982 Bangladesh National Drug Policy (Chowdhury 1995a), which is discussed later in this chapter.

This initiative to find alternative ways to meet primary health needs was reiterated in the 28th World Health Assembly in 1975. At the assembly, the goal of "Health for All by the Year 2000" was initially proposed by Dr. Halfdan T. Mahler, the WHO President at the time (Cueto 2004). Two years later at the Thirtieth World Health Assembly in 1977, experts on various systems of traditional medicine gathered together to create a plan to promote and develop traditional medicine. In this meeting, they acknowledged previous studies that WHO had conducted on the use of traditional systems of medicine. For example, during the 1950s, WHO and UNICEF, in cooperation with the Filipino government, provided biomedical training for local healers who assisted in birth. The idea was to integrate these traditional healers into the formal national healthcare system in order to reduce maternal and infant mortality in the Philippines (Kadetz 2011). Mao's integration of barefoot doctors into the national healthcare system in China and India's integration of Ayurveda and Unani into its national healthcare system were models that WHO encouraged other countries to follow (Pordié 2010). In 1978, WHO adopted its vision of integrating traditional medicine and its practitioners into national health systems by arguing that "effective integration, like that of the Chinese experience, entails a synthesis of the merits of both the traditional and the so-called "Western" or modern systems of medicine through the application of modern scientific knowledge and techniques" (WHO 1978b, 16). According to Kadetz (2011), the conflation of the Indian and Chinese examples by the WHO was problematic because in China the integration was forced in a top-down manner, whereas in India the traditional practitioners used their political power and agency to gain government support and recognition.

The Declaration of Alma-Ata

The idea of prioritizing primary health care was also carried over into the Declaration of Alma-Ata in 1978, ratified by 134 countries during a WHO and UNICEF sponsored international conference on primary health care in the former Soviet Union. This declaration embraced the concept of primary health care (PHC) and focused mainly on the use of "appropriate technology," the opposition to medical elitism, and the use of health as a tool for economic development. Health was expressed as a fundamental human right and primary health was defined as:

> essential health care based on practical, scientifically sound and socially acceptable methods and technology made universally accessible to individuals and families in the community through their full participation and at a cost that the community and country can afford to maintain at every stage of their development in the spirit of self-reliance and self-determination.
>
> (Declaration of Alma-Ata 1978)

The Declaration of Alma-Ata discouraged a top-down approach to health. Individual nations were encouraged to train lay health personnel and to involve local communities in healthcare provision. This declaration outlined advice to incorporate primary healthcare into national systems through the formulation of national policies, strategies, and plans of action. In order to do this, countries were called on to better and more fully use their national resources. Countries needed to rely:

> at local and referral levels, on health workers, including physicians, nurses, midwives, auxiliaries and community workers as applicable, as well as traditional practitioners as needed, suitably trained socially and technically to work as a health team and to respond to the expressed health needs of the community.
>
> (Declaration of Alma-Ata 1978)

A country's national resources included both traditional practitioners as well as traditional medicines. In order to meet the development goal of "Health for All by the Year 2000," the WHO promoted traditional healers as a resource to make up for the shortage in national healthcare delivery systems (Langwick 2008). The basic tenets of WHO's (2002) recommendations on traditional medicine were updated in 2002 as the "WHO Traditional Medicine Strategy 2002–2005".

Bangladesh's Second Five Year Plan

The influence of the Declaration of Alma-Ata was reflected in Bangladesh's Second Five Year Plan (1981–1985). One goal of this plan was to create a health system based on a regionalized healthcare model whereby the rural services would be the referral institutions to the more specialized urban healthcare system. Through this system, the state sought to advance family planning services, control communicable diseases, and expand immunization. The plan supported creating a steady supply and circulation of essential drugs by establishing new manufacturing plants in Bangladesh to make biomedical drugs.

In line with WHO recommendations, the Bangladesh government also expressed a renewed interest in integrating traditional medicine (at that time termed "indigenous medicine and homeopathy") into the modern healthcare system.[7] The emphasis in Bangladesh's Second Five Year Plan (SFYP) was to "develop them [indigenous systems] on proper scientific lines and to integrate them into the Primary Heath Care System" (Planning Commission 1983, 334). The language of science is similar to that used in colonial policies in British India, as much as it mirrored the post-Alma-Ata WHO policies. This development was to be carried out through the establishment a degree hospital with a 100-bed hospital in Dhaka for each of the systems[8] of traditional medicine. In 1983, in accordance with the SFYP, the government established the Bangladesh Unani and Ayurvedic Degree College and the hospital clinic described at the beginning of this chapter (Osman 2004, 245). The research goals at these institutions were to include scientific research on the efficacy of traditional medicine in the treatment of common diseases, with the aim of finding diseases that can be treated more effectively by traditional medicine than by allopathic medicine.

Traditional systems of medicine (Unani, Ayurveda, and homeopathy) in the SFYP are acknowledged as "deeply rooted in the country" (Planning Commission 1983, 334) because traditional practitioners treat a large segment of the population. Also, the main providers of healthcare in Bangladesh and South Asia have historically been practitioners of traditional medicine, especially in rural areas. A general familiarity with traditional practices of medicine was described as a possible benefit to public health. Communities that were not within reach of biomedical healthcare could have easier access to traditional medicine. In addition to the benefit of integrating traditional practitioners as a source of healthcare workforce, the drugs used by traditional medical practices are seen in the SFYP as an untapped national resource:

> Indigenous/traditional medicine is based on locally available crude drugs mainly herbs and plants. Bangladesh is rich in medicinal herbs and plants but no attempt has so far been made to utilize these resources systematically and scientifically although the efficacy of some of the indigenous drugs in some disease areas is widely recognized. . . . These two systems have practically been providing the bulk of health coverage to the mass particularly in the rural areas.
>
> (Planning Commission 1983, 334).

The move for public health to include traditional medicine was not new, but in the SFYP there was an added emphasis on their use as a national resource.

Government and WHO recommendations also supported the training and mainstreaming of all types of traditional and folk practitioners into either Unani or Ayurvedic systems of medicine. Moreover, with the WHO-supported 1987 Indigenous Medicine Development Project, two-month training programs were instituted and conducted by both Unani and Ayurveda diploma colleges in Bangladesh. These courses were designed for lineage family practitioners who had been practicing for at least ten years or for practitioners who had studied with a recognized *hakim* or *kobiraj* (Alam 2007, 200). Although the traditions of medicine were fluid in practice, these courses structured and standardized traditional medical knowledge in Bangladesh. The new categories of traditional medical practitioners aligned with the official categories of medicine. Traditional healing practices that were not integrated into categories of Unani or Ayurvedic medicine were not legitimate in the eyes of the state. In some estimates, there were recently 30,000–80,000 untrained and thus unregistered traditional practitioners (Alam 2007, 209; Osman 2004, 97–98). The state registries leave out *kobiraj* and *hakim* who do not meet state standards for registration for the categories of Unani and Ayurveda. These two terms and "systems" of medicine are slippery in practice, but when practitioners who learned in practice take the exam and register, they have to choose whether they are officially a practitioner of Unani or Ayurvedic medicine. Mukharji describes how historically there were many non-textual practices of traditional medicine in Bengal. Thus, many of these traditions, like *Chandashi*, were not accounted for historically and were left out of the written record.

Hence, traditions that were perhaps of more recent origins and not based on ancient texts and clearly articulated first principles were either ignored altogether under the label of "superstition" or "quackery," or indeed sometimes subsumed under any of the major traditions such as ayurveda and unani.

(Mukharji 2006, 278)

Regulating traditional medicine in Bangladesh

Within Bangladesh, recommendations from international organizations influenced new modes of governance in the form of drug policies that affected the production and circulation of all types of medicines, including traditional medicines. Global institutions such as the WHO and WTO became important authorities in the pharmaceutical governance of healthcare around the world. An important part of the practice of public health was the idea that a transfer of technology from the West to the developing world was the solution for healthcare. One technology that played a key role in this process was drugs, or pharmaceuticals. As described earlier, the idea of access to essential medicines became a key component of "Health for All." In 1977, the WHO published its formal definition of essential medicines in the *WHO Technical Report 615: The Selection of Essential Drugs.* This report promoted the use of generic drugs over brand names, single agents over combinations, drugs for common conditions over rare diseases, and safe and efficacious older drugs over new drugs (Greene 2011, 18). The vision of the list was largely based on guidelines published by Daniel Azarnoff, a professor of medicine in the United States and a member of the special WHO steering committee of doctors and public health experts. Azarnoff viewed essential drugs as a cost-effective and utilitarian way to improve public health by restricting the availability of drugs to only 150 medicines considered to be essential to the health of the majority of the public (Greene 2011; Chowdhury 1995b). This list of essential medications was only a model, and individual regions and countries were supposed to come up with their own national lists.

The connection between essential medicine and traditional medicine may not be significant in many parts of the world, but in Bangladesh, the WHO guidelines had an indirect effect on traditional medicines, which in turn affected their production and sales. The incorporation of essential medicines into the health policy meant that policymakers gave more scrutiny to health practices and decided to regulate all medicines, including – for the first time – the official registration of Unani and Ayurvedic medicines as drugs. These drug policies resulted in new legal categories of medicine that not all stakeholders agreed with. There was public debate about who was on the Drugs Committee, about banning the import of foreign-made medicines, and about which medicines (both allopathic and traditional) were considered essential and which ones were banned as being non-essential.

The WHO's guidelines for essential drugs lists and the increasing role and local critiques of the multinational pharmaceutical industry in the country eventually led to the creation of the 1982 Bangladesh Drugs Control Ordinance.[9] This ordinance was established under the rule of the military dictator, General Ershad. Influenced by recommendations from the WHO, the Drugs Ordinance established a list of 150 essential medicines, and at the same time prohibited the manufacture and sale of 1,742 pharmaceutical products in Bangladesh. Most of the banned products were tonics, lozenges, creams, and some liquid preparations with alcohol. The banned drugs were placed in three different categories:

- **Schedule I:** The committee recommended immediate stoppage of production of drugs listed in Schedule I. These drugs were to be collected from pharmacies and destroyed within three months of the acceptance of the report.
- **Schedule II:** Drugs in this category were to be reformulated within six months on the basis of the guidelines suggested by the committee.
- **Schedule III:** A maximum of nine months was allowed for utilization of Schedule III drugs.

(Chowdhury 1995b, 58)

The National Drug Policy also prohibited the importation of raw materials for Schedule I and II drugs. The ordinance included measures to protect local manufacturers, and no drug imports were allowed from multinational manufacturers if they were already being produced locally. These measures were a response to public criticism of multinational pharmaceutical companies in Bangladesh. This was significant in the early 1980s because at that time eight multinational pharmaceutical companies manufactured approximately 75% of all pharmaceuticals (in value) (Chowdhury 1995b).

The Drugs Control Ordinance also controlled all registrations, production, distribution/circulation, sale, export, and import processes in Bangladesh ("New Nat'l Drug Policy Handed over to PM" 2005). Zafrullah Chowdhury, considered a freedom fighter for Bangladesh independence and a grassroots activist, had an important role in the formulation of the policy. Reich argues that this policy articulated a vision of self-reliance and priority provision of basic national needs and an attitude of proud defiance against the multinationals – a stance of economic nationalism. Finally, the policy generated legitimacy through its support of international agencies and NGOs (Reich 1994, 133).

The Bangladesh Medical Association, local and international pharmaceutical firms, and foreign governments all opposed this new drug policy. Foreign governments argued that the new policy would discourage private investors from investing in Bangladesh. These governments had a lot of influence on polices in Bangladesh since, at the time, about 80% of Bangladesh's funds for development came from foreign aid. The Bangladesh government gradually allowed some banned products to be manufactured locally, and by 1986, most local pharmaceutical companies supported the drug policy. There was also a shift in the ownership of local pharmaceutical companies. In 1981, multinational corporations were the top five sellers of pharmaceuticals in Bangladesh, but by 1991, the top three pharmaceutical firms were all Bangladeshi-owned (Reich 1994). Reflecting on WHO's non-response to Bangladesh's National Drug Policy, Zafrullah

Chowdhury (1995a, 134) critiqued WHO's unwillingness to defend the new drug policy even though it was based on WHO recommendations, a contradiction he described as "a political decision *not* to act" (emphasis in original).

The Bangladesh Drugs Control Ordinance of 1982 was also significant because it was the first regulatory control over traditional medicine in Bangladesh and included the development of registration criteria. The pre-existing Drugs Act of 1940, created under British colonial rule, cursorily mentioned Unani and Ayurvedic medicines, but did not consider them drugs. Later amendments to the original act were concerned with the regulation of traditional medicines.[10] The new policy also called for creating a National Drug Control Laboratory which "would develop appropriate standards and specifications for Unani and Ayurvedic drugs" (Chowdhury 1995b, 60). During the 1980s, a time when the nation was liberalizing its economy, there was a rise in the number of local manufacturers of traditional medicines. Policymakers and members of the medical community were alarmed because they thought that there was not adequate control of traditional medicines. They also claimed that traditional drug manufacturers were increasing their sales by producing allopathic medicines banned by the ordinance. The resulting 1982 National Drug Policy included many new regulations to govern the drug industry through the monitoring and registration of traditional drug manufacturers (Chowdhury 1995b).

For example, officials found that traditional drug manufacturers were making medicines with greater than the allowed 5% of alcohol, a sensitive topic in Bangladesh, a predominantly Muslim country where the sale of alcohol is illegal. Traditional drug manufacturers, upset with this new alcohol policy, argued that some Ayurvedic medicines needed a higher percentage of alcohol to be effective. At the same time, Ayurvedic manufacturers criticized the government for interfering in the religious affairs of the Hindu community because the drugs were made in accordance to Hindu religious books. However, their claim was proven to be unfounded because most of the owners of Ayurvedic manufacturers in Bangladesh were actually Muslim and not Hindu (since many Hindu owners had fled the country). Also, an expert committee, consisting of doctors, policymakers, and practitioners of traditional medicines, was formed on 27 February 1983 to investigate accusations that some Ayurvedic and Unani manufacturers were producing banned allopathic medicines under the guise of traditional medicine. They also found that more than 25,000 brands were on the market at that time, and recommended only 431 medicines. In response, medicines of Schedule III and some traditional medicines were given an extension until 30 June 1984 to comply with the law (Chowdhury 1995b, 86–87).

The new official medical categories also determined the distribution of medicines in the space of pharmacy shops. Some traditional medicine manufacturers gave me copies of government circulars published in newspapers that they had archived in their own company files. One company owner explained that the government had first mandated that traditional medicines were to be sold in companies' own showrooms, but after pressure from the Unani and Ayurvedic Manufacturing Association, the government changed its stance. The first government circular was issued just two years after the 1982 Drugs Ordinance. This circular, posted on 15 August 1984, specifically allowed the legal sale of licensed categories of traditional medicine in allopathic pharmacies:

> From now on all licensed allopathic drug stores, retail and wholesale, can sell, store, and distribute organic, inorganic, and other drugs composed with them as long as they are registered and licensed Unani, Ayurvedic, and Homeopathic, and biochemic medicines.
>
> Those shops with allopathic licenses must have separate cabinet/shelves to store different kinds of medicines and they must be labeled as Unani, Ayurvedic, Homeopathic, or Biochemic.

The legal acceptance of licensed traditional medicines in allopathic pharmacies shows that allopathic and traditional medicines were both considered drugs to be regulated in the eyes of the state, but they were not seen as commensurate. Licensed traditional medicines were not given an equivalent status in pharmacy shops because they had to be marked as different and kept in physically separate spaces in the shops. Although the government circular about traditional medicine is a legal statement, many local people did not know the actual law. Therefore, some shop owners and traditional drug-making companies kept a paper copy of these circulars, cut out from newspapers, to ward off government officials or other people who did not know the law and tried to shut down their shops and harass them. The pharmacy shops were often raided and shopkeepers told that the traditional medicines were *oboighanik* or unscientific. The symbolic role of medicines in pharmacies is seen in other Asian contexts. In her work on the politics of herbal medicine cabinets in pharmacy shops in 1970s Korea, Ma (2010, 368) argues that these cabinets came to stand for a material and semiotic link between traditionalists and Western pharmacists. In Bangladesh, there were no cabinets at play, but in both situations, the authority of how to regulate the space of the pharmacy shop was contested because traditional medicines were symbolically linked to pre-modern and non-scientific ideas.

The 1982 Drug Ordinance created an official category for traditional medicines that gave them recognition and made them legally equivalent to allopathic medicine. However, this symbolic categorization did not always mean that representatives of traditional medical practices had equal power and representation in practice. Ten years after the drug ordinance, controversies emerged over the degree of control that the government had in the circulation of both allopathic and traditional medicines. Manufacturers of traditional medicine felt that the government was partial to influence from allopathic drug companies who were intent on monopolizing the national market. In 1992, the Ministry of Health and Family Welfare formed a new committee to prepare and amend the 1982 National Drug Policy. Because representatives from traditional medicine were not included in this committee, the Unani, Ayurvedic, and homeopathic manufacturing committees filed a lawsuit against the Bangladesh government. Even after the Ministry of Health had given assurances that the Unani, Ayurvedic, and homeopathic manufacturers would be represented in the new drug committee, they were not only left out of the Ministry of Health meeting, but also during the meeting it was decided that "Unani, Ayurved [sic], and Homeopathic medicines [were] not modern medicines and as such they did not feel it necessary to include their representatives in the said committee" (Writ Petition No. 3892 1992, 4). But Unani, Ayurvedic, and homeopathic medicines *were* included as drugs in the National Drug Ordinance of 1982. In their lawsuit in 1992, the traditional drug manufacturers alleged that the new review committee was breaking the law and violating the Constitution by *not* including them. In order to bolster the legitimacy of traditional medicine, the lawsuit included mention of WHO's recognition and support for the development of traditional medicine in countries around the world, including Bangladesh. The lawsuit also alleged that allopathic manufacturers had influenced the government's decision in order to retain their monopoly in the business of medicine (McNamara 2014). The traditional medicinal manufacturing committees won their lawsuit against the government six years later, in 1998, which gave them an official role in the governance of public health and medicines in Bangladesh.

A decade later, in 2008, representatives of formal practices of Unani and Ayurveda, from the colleges of traditional medicine and manufacturers of traditional medicine, used the symbolic reference to WHO again to legitimate their practices to the state. They organized meetings in the lead up to the 2008 national elections. During these events, they expressed grievances about the lack of government support for the traditional medicine sector and publicly made demands from the government. Themes of the meeting were the integration of medical practices and the cooperation between various stakeholders present. Attendees spoke about healthcare as a "right" and mentioned the role of traditional medicine in "Healthcare for All," a slogan of the WHO that was developed at the 1978 WHO conference in Alma-Ata.

Markets, patents, and herbal medicines

The increasing global market and circulation of herbal medicines has created additional dilemmas for the governance of traditional and herbal medicines in Bangladesh. WTO policies related to patent protection for biomedical drugs embroiled allopathic drug companies in controversy with traditional drug-making companies again. This time, a new category of herbal medicines was the controversial wedge between the two sides. This controversy was related to the creation of a new official category of herbal medicine and a 2016 WTO patent deadline.

In Bangladesh, most herbal pharmaceuticals are Unani or Ayurvedic medicines. Once made by local practitioners, these medicines are increasingly mass-produced by drug-manufacturing companies. Nowadays, new types of herbal medicines that are not Ayurvedic or Unani medicines are circulated, sold, and produced in Bangladesh. The Bangladeshi state has taken steps to encourage the growth of the herbal pharmaceutical industry. In 2004, it set up a business promotion council for the herbal pharmaceutical sector. According to the Export Promotion Bureau

(EPB), 136 homeopathic medicines, 52 Unani medicines, 52 Ayurvedic medicines, and 48 herbal plants are included on the list of exportable items (Haque 2004). In 2004, the herbal pharmaceutical market in Bangladesh was estimated to be around Tk 100 crore (over USD $17 million in 2004). Herbal companies like Hamdard were urging this sector in Bangladesh to reach a projected growth of Tk 1,000 crore[11] (Palma 2005). Herbal pharmaceutical companies in Bangladesh are eager to become a bigger player in the global market for herbal medicines, which the WHO reported was over USD 60 billion annually in 2005 and growing steadily (Tilburt and Kaptchuk 2008).[12]

Part of the government promotion of herbal medicines includes a call to increase the cultivation of herbal plants in Bangladesh, in order to become less dependent on imports of raw materials. A rising number of farmers are interested in growing medicinal plants in order to supply the burgeoning herbal pharmaceutical industry. Many of these farmers live in an area called Kholabaria, in the western part of the country. The government is encouraging the development of the herbal pharmaceutical industry as a potential boost to the national economy. They suggest that if Bangladesh could increase its production of herbal medicine, the country could meet national demands, earn a large amount of foreign currency, and create income for the rural people who would cultivate the medicinal plants. Yet some farmers worry about their future sales because there is no guarantee that the pharmaceutical companies will actually buy their plants. They don't have direct access to the companies, but sell their medicinal plants through a complex network of traders and middle-men (McNamara 2016).

The increasing global circulation of herbal medicines has resulted in new drug policies in Bangladesh that aim to govern the definition, use, and circulation of herbal medicines locally. In Bangladesh, a new National Drug Policy was formulated in 2005. This policy officially defined herbal medicine and partially took advantage of the trade-related aspects of intellectual property rights (TRIPS) rule that allowed least developed countries (LDCs) to produce essential drugs without requiring patents until 2016. TRIPS is the basis for the global intellectual property regime that is regulated by the WTO and was signed into agreement in 1993 by two-thirds of the world's nations. TRIPS protects the patents and processes of drugs for 20 years after they are registered. All counties that are members of the WTO are required to introduce or upgrade a standardized level of protection of drug patents. A country's state of "development" determined their deadline for complying with the WTO requirements. Economically advanced countries had to comply by 1996, "developing countries" by 2000 (some extended until 2005), and 49 countries considered the "least developed" had until 2016 to pass legislation requiring patent protection for allopathic pharmaceuticals and agricultural chemicals (Foreman 2002).

Because Bangladesh is categorized as a "least developed country," drug manufacturers in Bangladesh could manufacture generic allopathic pharmaceuticals until 2016 without purchasing patents. Bangladesh is the only LDC country, out of 49, to have a manufacturing base ("Govt Plans to Upgrade Drug Admin Directorate, Move to Help Local Cos to Tap Overseas Markets" 2004). This put

Bangladesh in a unique position to take advantage of TRIPS by manufacturing many biomedical drugs locally and selling them both locally and internationally. In addition, the 2005 drug policy called for more collaboration between universities, research institutes, and manufacturers with the goal of strengthening the pharmaceutical industry (Zannat 2007). The private sector was frustrated with the government delay in improving the Bangladesh Drug Testing Laboratory, which was not recognized by some European nations, and made plans for opening its own independent standardized testing laboratory ("Govt Plans to Upgrade Drug Admin Directorate, Move to Help Local Cos to Tap Overseas Markets" 2004). There was also extensive public criticism of this new policy because the repeal of the price control mechanism made many drugs unaffordable at the local level (Zannat 2007).

Herbal medicines controversies

The 2005 Bangladesh National Drug Policy also mentioned a new category of Herbal medicines for the first time in its description of registration criteria:

i) As a general principle, registration for manufacture, import, and sale of combination drugs other than those of Unani, Ayurvedic & other Herbal preparations, vitamins and nutritional preparations should not be allowed in the country. However, combinations like vitamins, nutritional preparations and other drugs which are therapeutically useful and are registered in the developed countries could be considered for registration. . .

iv) Any Unani, Ayurvedic, or other Herbal drugs included in the official Formulary of other countries, [or] is considered essential and useful by the National Unani, Ayurvedic & Herbal Formulary Committee, may be granted registration for manufacture and sale in the country. National Unani, Ayurvedic, & other Herbal Formulary Committee will be constituted by the government with the experts on Unani, Ayurvedic and Herbal drugs.

(Ministry of Health and Family Welfare 2005, 5)

The inclusion of herbal medicine in this policy was significant because it broadened the official definition of herbal drugs to include herbal medicines outside of Unani and Ayurvedic medical traditions. Official government drug formularies are important in Bangladesh because all drugs manufactured in the country must be listed in one of the formularies and companies need specific manufacturing licenses to produce drugs.[13] The new policy also sought to improve the quality control of the production of Unani and Ayurvedic medicines. All manufactured medicines were required to follow the guidelines of good manufacturing practices (GMP), as recommended by the WHO. According to the WHO, GMP are "a system for ensuring that products are consistently produced and controlled according to quality standards. It is designed to minimize the risks involved in any pharmaceutical production that cannot be eliminated through testing the final product."[14] If a product meets the standards for base materials, equipment, documentation, premises, processes, training, and personal staff hygiene, then a GMP certificate is given to its manufacturing facility. The WHO does not carry out any of these inspections, although it does set the standards for GMP and current good manufacturing practices (cGMP). In Bangladesh, the Directorate of Drug Administration (DDA) or international organizations, such as UNICEF, issue the GMP certificates (The World Bank 2008).[15] Regulating drug production and complying with the WHO's GMP encourages drug-manufacturing companies to standardize their drug production. This standardization acts as a type of symbolic capital that gives them eligibility to enter the global marketplace. These standards and classification influence the governance of traditional medicine and become powerful symbols which help in the global circulation and sales of medicine (Petryna and et al. 2006).

In 2007, the tensions between traditional drug-making companies and the government peaked when the Bangladesh Drug Directorate expanded on the 2005 drug policy to create a new herbal drug-manufacturing license to regulate the increasing interest in and demands for new types of herbal medicine. A group of traditional drug manufacturers filed a new lawsuit against the government, in an attempt to rescind the new herbal drug-manufacturing license. The previous lawsuit (in 1992) mentioned earlier in the chapter, won by traditional drug manufacturers, had alleged that allopathic companies were influencing the government against traditional medicines. This time, traditional drug companies claimed that allopathic companies had pushed the Bangladesh government into issuing a new herbal drug-manufacturing license. Most Unani and Ayurvedic companies were not interested in applying for the new herbal drug license because they manufactured their medicines after obtaining the already existing Ayurvedic or Unani drug license. Instead, it was mostly allopathic companies[16] that were interested in manufacturing new herbal medicines, in light of the global circulation and popularity of natural herbal medicines. Because of WTO patent policies and the end of the TRIPS agreement in 2016, allopathic companies in Bangladesh did not know how long they could use free patents to produce their allopathic medicines. This uncertainty had a strong influence on why allopathic companies had a new interest in new "herbal" medicines that weren't Unani or Ayurvedic – in order to expand their uncertain market (McNamara 2014). In November 2015, WTO members agreed to extend the TRIPS deadline until 2033 for LDCs, which includes Bangladesh.[17] We will have to wait and see if this extension affects allopathic companies' interest in making herbal medicines.

Pharmaceutical manufacturers in Bangladesh meet approximately 97% of local demand (Ahmed 2015). In addition, from 2001–2008, the number of foreign countries receiving pharmaceutical exports from Bangladesh increased from 17 to 71 (Table 1.1). The revenue from these exports increased from Tk 32 crore (USD 5.8 million[18]) in 2001 to Tk 541 crore (USD 85.3 million[19]) in 2005 (National Drug Policy Announced, New Policy Discourages Import of Drug 2005). As seen in the Table 1.1, the revenue from medicines (this includes all types of medicines) exported as a finished product from Bangladesh in this seven-year time span increased by tenfold. In adherence with the Doha Declaration of 2001, countries abiding by TRIPS

Table 1.1 Drug-manufacturing export statistics for Bangladesh (2008 government data)[1]

Year	Finished Products (in million taka)	Raw Materials (in million taka)	No. of Export Countries
2001	311.8	11	17
2002	406.91	43	32
2003	545.46	87.32	51
2004	1400	138.97	62
2005	1421	147.57	67
2006	2519.98	143.41	61
2007	2347.2	130.31	67
2008	~3000	146.11	71

[1] The data in this chart were shared with me when I met with an official at the Bangladesh Directorate of Drugs in 2008.

regulations were given greater patent flexibility to circumvent patent rights. In Bangladesh, in accordance with this declaration, the Department of Patents, Designs and Trademarks suspended the patenting of pharmaceuticals from 2008 until the beginning of January 2016. In the previous ten years, nearly half the patents filed were for pharmaceuticals – filed mostly by foreigners and multinational companies (Azam and Richardson 2010). The TRIPS agreement did make plant varieties patentable for commercial use. However, most traditional drug-making companies were not aware of the patent system or intellectual property rights. As of 2007, only three herbal pharmaceutical companies had patented products. One of these companies is Hamdard, the largest Unani company, which patented a product for the global market. According to the Patent Office, up until 2007 there had been no patent applications for innovation of herbal products in Bangladesh (Alam 2007).

There has not been a big a push for global patents in Bangladesh, although some herbal companies are becoming more interested in the global market and in protecting their brands and formulations of medicine. For example, one company in particular that makes Unani, Ayurvedic, and new herbal medicines was particularly concerned about the fake medicines that imitated its brands and tradenames. The director of this company sent scouts out into the local bazaars to find these *nokol oshudh*, or fake medicines, because he was concerned that these medicines would give a *bodnam*, or bad name, to his company medicines and ruin its reputation. He said that he often changes the packaging and labels of his medicines to make them harder to imitate. This same director had an interest in exporting medicines to new markets in predominantly Muslim countries in Africa. These countries have fewer regulatory controls on imports, and he was appealing to their shared Muslim identity to market his medicines. He even showed me an email printout with a recent order to export over USD 1 million worth of his products to countries in West Africa.

In August of 2009, I asked an official from the Bangladesh Drug Administration about the lawsuit involving the new herbal manufacturing license. He did not seem very concerned about the lawsuit, or that it would change the way herbal drug-manufacturing licenses are issued. He explained that the government created the new herbal drug license in 2007 because some herbal medicines exist outside of Unani and Ayurveda and are relevant to the modern world because people still consume them. Herbal drug licenses could be granted because the board had agreed on a definition of herbal medicine, testing criteria, and reference books for manufacturers to use. He told me that the Drug Administration had come up with the following list of reference books for the new herbal medicines:

- *The American Botanical Council Clinical Guide of Herbs*
- *American Herbal Pharmacopeia & Therapeutic Compendium*
- *British Herbal Pharmacopeia*
- *The Complete German Commission E Monographs Therapeutic Guide to Herbal Medicines*
- *E/S/C/O/P Monographs, The Scientific Foundation for Herbal Products*
- *Herbal Drugs and Phytopharmaceuticals*
- *Mosby's Drug Consultant*
- *PDR for Herbal Medicines*
- *WHO Monographs on Selected Medicinal Plants*

This list shows how the Bangladesh Drug Administration is drawing from a much wider range of sources, including WHO publications, to officially define what drugs could be produced under the new herbal drug license (McNamara 2014). Previously, the official references were only the national pharmacopeias for Unani and Ayurvedic medicines. Herbal drug-manufacturing licenses could now be granted, because both testing criteria and a set of reference books for the drug manufacturers had a list of "herbal" pharmacopeias to draw from to produce new types of herbal medicines. The Drug Administration hoped that the drug-manufacturing sector would grow in a more disciplined manner with the finalization of these criteria. However, it is not always clear which pharmacopeias drug-manufacturing companies choose to draw from. At that time, only four companies were registered with herbal drug-manufacturing licenses, but more than 20 companies had applied for it and were awaiting approval (Parvez 2009).

A lawyer who was working on behalf of the Unani and Ayurvedic drug manufacturers which had filed the new lawsuit against the government, contesting the herbal drug-manufacturing license, explained the premise of the case. Because there was no recognition of the system of new herbal medicine in the existing law, the new lawsuit argued that the new herbal manufacturing licenses were illegal and should never have been issued in the first place. This court case was not yet settled at that time, so the lawyer did not want to share all the details of the case with me. Representatives from some Unani and Ayurvedic companies expressed a certain moral indignation with the new herbal manufacturing license, and claimed that the new herbal medicines had no relation to the history and traditions of Bangladesh. One Unani manufacturer told me that the new herbal medicines are less trustworthy because there is no standardized formulary or pharmacopeia, and access to the internet encourages even more variance. Traditional manufacturers feared that the economic value of drugs would create more competition in the market for herbal medicines. Interestingly, there were a few Unani and Ayurvedic companies who were not opposed to the new herbal drug-manufacturing license, and registered for it in order to expand the types of medicines they could make and sell.

The lack of oversight and manpower challenges Bangladesh's ability to enforce regulations on both traditional and herbal medicine. These failures to enforce regulations for all types of medicines is a regular feature of local newspaper articles that lament the sales of fake medicines or highlight raids on unregistered drug-manufacturing operations.

Conclusion

WHO guidelines were translated into new forms of national governance that affected the production and circulation of traditional and herbal medicines in Bangladesh. When I asked Asif, the WHO employee in Bangladesh mentioned at the beginning of this chapter, if WHO had any stance on the debates about herbal medicine in Bangladesh, he said that WHO does not have one individual opinion on these things because the plans for different countries are tentative and experts at global, regional, and national levels set the priorities for governance:

> if we say WHO strategy on the use of herbal medicine, in fact there is no one person in the WHO who just sat down and put that strategy. . . . So WHO does not have any stand on these things because traditional medicine means . . . it evolved historically, the degree of efficacy and efficiency, WHO is not in a position to validate everything. WHO supports countries to make the best of what is available in their countries and then to generate evidence and to collect information about how much this medicine is effective and safe.

WHO recommendations and strategies have influenced the institutionalization of traditional medicine in Bangladesh, but often for contradictory purposes. In their first lawsuit against the government, stakeholders in traditional medicine mentioned the WHO as a *de facto* witness and as a legitimate source of knowledge and evidence. The WHO's authority as a global institution was a symbolic bolster for the claims of the traditional drug manufacturers. However, the state also calls on the authority of the WHO to enact policies that are sometimes against the interests of traditional drug manufacturers. WHO gives technical advice but does not establish institutions. In Bangladesh, this advice consisted of helping to establish the curriculum for the schools, guidelines for ensuring the efficacy of medicines, good manufacturing practices, and quality control. As Asif said, "From the WHO side, we just help them in getting resource persons . . . because we don't have experts in this area. We bring resource persons [together] as a working group to work on it."

Jasonoff (2008, 776) uses the term "boundary organization" to describe "bodies such as expert advisory committees whose primary function is to maintain a clear demarcation between the authority of experts and political decision makers."

WHO acts as a boundary organization when it coproduces the mutual interests of both the government and representatives from the formal traditional medicine sector. For example, WHO also helped to establish and update the list of essential medicines for Bangladesh. When I asked if the list included bans on medicines, Asif explained that the list only says what medicines are essential, not what medicines are available or what medicines should be banned. He emphasized that the list was not a policy. Referring to treatment guidelines Asif emphasized that "it has nothing to do with government. This is technical. Technical material. . . . What's available in a country to do this and that. It has nothing to do with government policy. It is purely technical, there is no policy in it." This explanation exemplifies the attempt of the WHO, as a boundary organization, to remain neutral. Greene describes the dilemma inherent in the WHO:

> On the one hand, the framers of the early WHO took pains to emphasize that the organization was *not* a supranational regulatory agency, but instead represented a site for expediting the *normative* harmonization of its constituents sovereign regulatory agencies through an iterative process of convening expert conferences and publishing consensus standards. At the same time, the WHO was also understood to have an explicitly *activist* set of responsibilities in the field of international epidemiological intelligence and the international coordination of responses to epidemic disease threats.
>
> (Greene 2011, 15, emphasis in original)

Interestingly, the essential medicine concept has now expanded to include essential Unani and Ayurvedic medicines in the newest Bangladeshi drug policy, developed in 2016. The government enacted this policy to comply with international standards and it is considered the first integrated drug policy of its kind in South and Southeast Asia. The policy states that no drugs can be bought without prescriptions, with the exception of 39 allopathic, 23 Ayurvedic and 48 Unani medicines (Tusher 2016). We will have to wait and see how these new categories of essential medicines will affect their production and consumption.

The debates around medicines in Bangladesh reveal the struggles on global, regional, and national scales to regulate and standardize traditional medicine for institutions, for drugs lists and manufacturing purposes, and for the global market. The categories of medicine fracture when the processes of institutionalization are not agreed upon, or when categories of medicine erase or reify differences, such as between allopathic, traditional, and herbal medicine. We can see this in the attempt to distinguish herbal medicine as a category different from traditional medicine, even though traditional medicines are considered to be herbal medicines. The process of attempting to make medicines equal in measure, or commensurate, is fraught with inherent contradictions. In this process, different qualities are transformed into a common metric, thereby influencing how we categorize and make sense of the world (Espeland and Stevens 1998; Bowker and Star 1999). The process of commensuration is often invisible when systems or categories are taken for granted, as in the understanding of traditional medicine.

Traditional medicine was understood to be that which was not biomedicine, until a new category of herbal medicine came into play.

The process of commensuration creates homogeneity and also affects how cultural, legal, and economic values of different types of medicines are conceptualized. The Bangladeshi state values the economic potential of herbal and traditional medicines, but at the same time is reticent to view them as important as allopathic medicines. Meanwhile, some traditional drug-making companies feel like the category of new herbal medicines threatens the cultural and national values of traditional medicines, with an added fear that the new herbal medicines will diminish the economic value of traditional medicines. The as-yet-unanswered questions are how much companies and consumers in Bangladesh value local herbal and traditional medicines over global herbal medicines, and how to even recognize the difference between the two. Global and regional markets, international organizations such as WHO and WTO, and local actors use different measures to determine how and what drugs and medical practices are commensurate. The lawsuits discussed in this chapter demonstrate how the power to determine the legal value of different categories of medicines is uneven and increasingly influenced by their circulation in the global market. Sometimes, issues of equivalence and commensurability are not contested until they are codified in the law which "provides an envelope of social order within which new epistemic constructs and technological objects are constantly fitted out with recognizable meanings and normative implications" (Jasonoff 2008, 764). However, the constructs and objects of the law do not always represent the multiple cultural values held in a society. In Bangladesh, many forms of herbal medicine that are not mass-produced are "othered" by the more commodified and standardized Unani and Ayurvedic medicines. This process is intensified as pharmaceutical companies compete in the same markets and strive to maintain good reputations for their brands. Although there is vast knowledge about herbal and traditional medicines in Bangladesh, there is still a reliance on the expertise of international organizations, like the WHO and WTO, to govern and validate this knowledge with the increasing circulation of medicines around the world.

Notes

1 An American Institute of Bangladesh Studies Junior Research Fellowship and a Fulbright-Hays Doctoral Dissertation Research Abroad fellowship supported the research for this chapter. An earlier version of this chapter was awarded the Charles Leslie Junior Scholar Award for Best Presentation at the 2013 International Congress on Traditional Asian Medicine. The discussion at this conference and the feedback from Céline Coderey and Laurent Pordié helped me to refine my argument. A postdoctoral fellowship at the Asia Research Institute at the National University of Singapore gave me the space and time to revise my paper into this chapter. I am indebted to the many practitioners and manufacturers of traditional medicine in Bangladesh who made my research possible and took the time to answer my many questions.

2 In order to protect their identities, all names of informants used in this chapter are pseudonyms.

3 The government defines traditional medicine as the formal practices of Unani, Ayurveda, and homeopathy.

4 Equivalent to around USD \$25,300 in 1978, which would be equivalent to approximately USD \$99,600 in 2017.

5 Although this was the first government degree college in traditional medicine, there were other private colleges of traditional medicine that were already established.

6 Bangladesh was East Pakistan until its independence in 1971.

7 Traditional medicine in Bangladesh was not as fully integrated into the public health system as it was in India due to decisions made by the Pakistani government after Partition in 1947. From 1947 until its independence in 1971, Bangladesh was governed by Pakistan as East Pakistan. The Chopra Report of 1948 encouraged the Indian government to synthesize traditional medicine with Western medicine and standardize the making of all medicines. Some practitioners of traditional medicine contested this idea of medical integration (Banerjee 2009, Wujastyk 2008). In Pakistan, there was even less support for traditional medicine because at the All-Pakistan Health Conference of 1951, the committee on indigenous medicines voted unanimously not to give state recognition to indigenous and homeopathic systems of medicine.

8 Unani, Ayurveda, and homeopathy.

9 Also called the National Drug Policy.

10 In India, the Drugs Amendment Act of 1964 added Unani and Ayurvedic medicines to the original 1940 Act. This amendment was to control the production and sales of these medicines (Srivastava 2010).

11 Approximately USD \$170 million in 2004.

12 www.who.int/bulletin/volumes/86/8/07-042820/en/ Accessed 3 April 2018.

13 In India, pharmaceutical companies conduct research to reformulate traditional drug formulas in the classic texts as new medicines that they register as trademarks under an "Ayurvedic Proprietary Medicine" label. This process is quicker and cheaper than applying for a patent (Pordié and Gaudillière 2014). In Bangladesh, after Unani and Ayurvedic drug manufacturers obtain a manufacturing license, they can legally produce medicines that are listed on one of the official national drug formularies. The Bangladesh Unani and Ayurvedic Board creates the formularies for traditional medicines. However, drug companies can register "tradenames" with the government Drug Administration. The National Formulary for Ayurvedic Medicine was created in 1992 and the National Formulary for Unani Medicine was created in 1993 (Alam 2007).

14 "GMP Questions and Answers." www.who.int/medicines/areas/quality_safety/quality_assurance/gmp/en/ Accessed 1 April 2018.

15 "Public and Private Sector Approaches to Improving Pharmaceutical Quality in Bangladesh." http://apps.who.int/medicinedocs/documents/s16761e/s16761e.pdf Accessed 1 April 2018.

16 However, some Unani and Ayurvedic companies later applied for an herbal manufacturing license to add to their products for sale – with an interest in increasing revenue.

17 "WTO members agree to extend drug patent exemption for poorest members." www.wto.org/english/news_e/news15_e/trip_06nov15_e.htm Accessed 2 April 2018.

18 Equivalent to USD \$8.2 million in 2017.

19 Equivalent to USD \$110.5 million in 2017.

References

Afdhal, Ahmad Fuad, and Robert L. Welsch. 1988. "The Rise of the Modern *Jamu* Industry in Indonesia: A Preliminary Overview." In *The Context of Medicines in Developing Countries: Studies in Pharmaceutical Anthropology*, edited by Sjaak van der Geest and Susan Reynolds Whyte, 149–172. Dordrecht: Kluwer.

Ahmed, Gazi Towhid. 2015. "Pharma Exports Rise on Growing Demand." *Daily Star*, March 8. www.thedailystar.net/pharma-exports-rise-on-growing-demand-17383. Accessed April 4, 2018.

Alam, Md. Jahangir. 2007. *Traditional Medicine in Bangladesh*. Dhaka: Asiatic Society of Bangladesh.

Azam, Mohammad M., and Kristy Richardson. 2010. "Pharmaceutical Patent Protection and Trips Challenges for Bangladesh: An Appraisal of Bangladesh's Patent Office and Department of Drug Administration." *Bond Law Review* 22 (2): 1–15.

Banerjee, Madhulika. 2009. *Power, Knowledge, Medicine: Ayurvedic Pharmaceuticals at Home and in the World*. New Delhi: Orient Blackswan.

Biehl, João. 2006. "Pharmaceutical Governance." In *Global Pharmaceuticals: Ethics, Markets, Practices*, edited by Adriana Petryna, Andrew Lakoff, and Arthur Kleinman, 206–239. Durham, NC: Duke University Press.

Bode, Maarten. 2002. "Indian Indigenous Pharmaceuticals: Tradition, Modernity, and Nature." In *Plural Medicine, Tradition and Modernity, 1800–2000*, edited by Waltraud Ernst. London: Routledge.

Bode, Maarten. 2008. *Taking Traditional Knowledge to the Market: The Modern Image of the Ayurvedic and Unani Industry, 1980–2000*. Hyderabad, India: Orient Longman.

Bowker, Geoffrey, and Susan Leigh Star. 1999. *Sorting Things Out: Classification and Its Consequences*. Boston: MIT Press.

Brown, Theodore M., Marcos Cueto, and Elizabeth Fee. 2006. "The World Health Organization and the Transition from 'International' to 'Global' Public Health." *American Journal of Public Health* 96 (1): 62–72.

Chowdhury, Zafrullah. 1995a. "Bangladesh: A Tough Battle for a National Drug Policy." *The Journal of the Dag Hammarskjold Foundation*. Development Dialogue (1): 96–146.

Chowdhury, Zafrullah. 1995b. *The Politics of Essential Drugs: The Makings of a Successful Health Strategy: Lessons from Bangladesh*. London: Zed Books.

Craig, Sienna, and Denise Glover. 2009. "Conservation, Cultivation, and Commodification of Medicinal Plants in the Greater Himalayan-Tibetan Plateau." *Asian Medicine* 5: 219–242.

Cueto, Marcos. 2004. "The Origins of Primary Health Care and Selective Primary Health Care." *American Journal of Public Health* 94 (11): 1864–1874.

Declaration of Alma-Ata. 1978. International Conference on Primary Health Care, Alma-Ata, USSR, September 6–12. www.who.int/publications/almaata_declaration_en.pdf.

Espeland, Wendy Nelson, and Mitchell L. Stevens. 1998. "Commensuration as a Social Process." *Annual Review of Sociology* 24 (1): 313–343.

Foreman, Martin. 2002. *Patents, Pills and Public Health: Can TRIPS Deliver?* London: The Panos Institute.

Gaudillière, Jean-Paul. 2014. "Herbalised Ayurveda? Reformulation, Plant Management and the 'Pharmaceticalisation' of Indian 'Traditional' Medicine." *Asian Medicine* 9: 171–205.

"Govt Plans to Upgrade Drug Admin Directorate, Move to Help Local Cos to Tap Overseas Markets." 2004. *The Daily Star*, Monday, November 22, 5 (177). http://archive. thedailystar.net/2004/11/22/d41122050154.htm

Greene, Jeremy A. 2011. "Making Medicines Essential: The Emergent Centrality of Pharmaceuticals in Global Health." *BioSocieties* 6 (1): 10–33.

Habib, S. Irfan, and Druv Raina. 2005. "Reinventing Traditional Medicine: Method, Institutional Change, and the Manufacture of Drugs and Medication in Late Colonial India." In *Asian Medicine and Globalization*, edited by Joseph Alter, 67–77. Philadelphia: University of Pennsylvania Press.

Haque, Zahidul. 2004. "Business Promotion Council for Herbal Medicine Sector on Cards." *The Daily Star* 5 (67). Dhaka.

Hardiman, David. 2009. "Indian Medical Indigeneity: From Nationalist Assertion to the Global Market." *Social History* 34 (3): 263–283.

Jasonoff, Sheila. 2008. "Making Order: Law and Science in Action." In *The Handbook of Science and Technology Studies*, edited by Edward J. Hackett et al., 761–786. Cambridge, MA: The Massachusetts Institute of Technology Press.

Kadetz, Paul. 2011. "Assumptions of Global Beneficence: Health Care Disparity, the WHO, and the Effects of Global Integrative Health Care Policy on Local Levels in the Philippines." *Biosocieties* 8: 88–105.

Langwick, Stacey A. 2008. "Articulate(d) Bodies: Traditional Medicine in a Tanzanian Hospital." *American Ethnologist* 35 (3): 428–439.

Ma, Eunjeong. 2010. "The Medicine Cabinet: Korean Medicine under Dispute." *East Asian Science, Technology and Society* 4: 367–382.

McNamara, Karen. 2014. "Establishing a Traditional Medicine Industry in Bangladesh." In *South Asia in the World: An Introduction*, edited by Susan Snow Wadley, 185–199. New York: M.E. Sharpe.

McNamara, Karen. 2016. "Medicinal Plants in Bangladesh: Planting Seeds of Care in the Weeds of Neoliberalism." In *Plants and Health: New Perspectives on the Health-Environment-Plant Nexus*, edited by E. A. Olson and J. R. Stepp, 67–91. Cham, Switzerland: Springer International Publishing.

Ministry of Health and Family Welfare. 2005. *National Drug Policy*. No. Public Health-1/Drug-22/2004/154. Government of Bangladesh, May 5.

Mukharji, Projit Bihari. 2006. "Going Beyond Elite Medical Traditions: The Case of Chandshi." *Asian Medicine: Tradition and Modernity* 2 (2): 277–291.

"National Drug Policy Announced, New Policy Discourages Import of Drug." 2005. *The Daily Star*, Thursday, August 18, 5 (437). http://archive.thedailystar.net/2005/08/18/d50818060667.htm

"New Nat'l Drug Policy Handed over to PM." 2005. *The Daily Star*, Monday, August 15, 5 (434). http://archive.thedailystar.net/2005/08/15/d50815060557.htm

Osman, Ferdous Arfina. 2004. *Policy Making in Bangladesh: A Study of the Health Policy Process*. Dhaka, Bangladesh: A H Development Publishing House.

Palma, Porimol. 2005. "Herbal medical system unregulated." *The Daily Star*, Monday, March 28. Dhaka, Bangladesh.

Parvez, Sohel. 2009. "Herbal Medicines get New Lease of Life." *The Daily Star*, July 11. Dhaka, Bangladesh.

Petryna, Adriana, Andrew Lakoff, and Arthur Kleinman. 2006. *Global Pharmaceuticals: Ethics, Markets, Practices*. Durham: Duke University Press.

Planning Commission. (1981). *Second Five Year Plan, 1981–1985*. Dhaka: Government of Bangladesh.

Pordié, Laurent. 2010. "The Politics of Therapeutic Evaluation in Asian Medicine." *Economic and Political Weekly*, May 1, 45 (18).

Pordié, Laurent, and Jean-Paul Gaudillière. 2014. "The Reformulation Regime in Drug Discovery: Revisiting Polyherbals and Property Rights in the Ayurvedic Industry." *East Asian Science, Technology and Society: An International Journal* 8 (1): 57–79.

Pordié, Laurent, and Anita Hardon. 2015. "Drugs' Stories and Itineraries: On the Making of Asian Industrial Medicines." *Anthropology & Medicine* 22 (1): 1–6.

Reich, Michael R. 1994. "Bangladesh Pharmaceutical Policy and Politics." *Health Policy and Planning* 9 (2): 130–143.

Srivastava, Lily. 2010. *Law and Medicine*. New Delhi: Universal Law Publishing Company.

The World Bank. 2008. "Public and Private Sector Approaches to Improving Pharmaceutical Quality in Bangladesh." *Bangladesh Development Series Paper* 23. Dhaka: Country Management Unit of the World Bank Office.

Tilburt, Jon C., and Ted J. Kaptchuk. 2008. "Herbal Medicine Research and Global Health: An Ethical Analysis." *Bulletin of the World Health Organization* 86 (8): 577–656.

Tusher, Hasan Jahid. 2016. "Updated Drug Policy Okayed by Cabinet." *The Daily Star*, December 20. Dhaka, Bangladesh.

Wahlberg, Ayo. 2008. "Pathways to Plausibility: When Herbs Become Pills." *Biosocieties* 3 (1): 37–56.

World Health Organization. 1977. No. 240, Part I Resolutions and Decisions Annexes. Thirtieth World Health Assembly. Geneva, May 2–19.

World Health Organization. 1978a. *Alma Ata Declaration*. Geneva: World Health Organization.

World Health Organization. 1978b. "The Promotion and Development of Traditional Medicine." *Technical Report Series 622*. Geneva: World Health Organization.

World Health Organization. 2002. *WHO Traditional Medicine Strategy 2002–2005*. Geneva: World Health Organization.

Writ Petition No. 3892. 1992. Supreme Court of Bangladesh. Dhaka, Bangladesh.

Wujastyk, Dominik. 2008. "The Evolution of Indian Government Policy on Ayurveda in the Twentieth Century." In *Modern and Global Ayurveda: Pluralism and Paradigms*, edited by Dagmar Wujastyk and Frederick M. Smith, 43–76. Albany: SUNY Press.

Zannat, Mahbuba. 2007. "Drug Policy Lacks Control over Price." *The Daily Star* 5 (1040). Dhaka, Bangladesh.

2 Science as a global governance and circulation tool?

The *Baekshuoh* disaster in South Korea

Eunjeong Ma

The 2015 Nobel Prize in physiology or medicine was awarded to the Chinese researcher who extracted the malaria-fighting compound artemisinin from the plant *Artemisia annua* (Callaway and Cyranoski 2015).[1] A curative effect of the herb is well recorded in traditional Chinese medicine, and the Nobel laureate was successful in isolating its biologically active ingredient and clarified how it worked to turn into a modern medicine. What made the discovery possible, according to Martha Hanson (2015), was to do with a quality named "medical bilingualism." Being medically bilingual refers to the ability to read and understand two different medical languages, their idiosyncratic histories, conceptual differences, and potential values for therapeutic interventions in the present. In her essay, Hanson (2015) speaks of an individual scholar whose expertise and practice has developed with the acquisition of two different medical traditions. Inspired by the notion, in this chapter, I address "coerced medical/pharmaceutical bilingualism." It refers to the non-voluntary fusion of two medical traditions when the government enforces the pharmaceutical industry to commercialize Korean medicines into health products.

As the consumption of health products such as dietary and nutritional supplements has increased since the 1980s (Han 1997), the South Korean government strategically supported science-based evidence as a governing instrument for the circulation of health products derived from traditional medicines. This regulatory move was also taking into consideration the international trend that employed intellectual property rights to utilize traditional knowledge and medicines. In a way, it was an administrative magic wand to enforce the crossings of two medical systems – Korean medicine and Western biomedicine – that have coexisted in public and private healthcare sectors in South Korea. Since the early 1990s, South Korean governments have sponsored the transformation of Korean (or herbal) medicines into commercial products, whether patented or not, under the rubric of modernization, scientization, and globalization (Jang 2001, 2011; Korea Drug Research Association 1992; INS 1998). Partly in reaction to the imminent global order grounded on intellectual property rights, the South Korean government turned to traditional Korean medicine and herbal products, viewing them as comparatively competitive market players in the near future. Pointing to such a successful case as artemisinin, the South Korean government stepped in to lure the domestic pharmaceutical industry and research institutes into translating herbal medicines into biomedical products (Ma 2015).

At one level, the government directed the domestic pharmaceutical industry and venture capitalists to reutilize traditional knowledge of medicine in order to establish the industrial system to mass-produce natural products that are marketable to domestic as well as international customers. In large part, a series of state-driven Korean medicines development projects led to the cultivation of biopharmaceutical industry, with an explosion of health-related products in the domestic market starting in the 1990s. At another level, the government's initiative to transform Korean medicines opened up a political and epistemic space where professional boundaries over the ownership of Korean medicines are renegotiated and redefined (Ma 2010, 2015). Representative professional organizations from Western biomedicine and Korean medicine have been claiming for a rightful ownership over commercialized Korean medicines, based on conflicting reasons. From the side of Korean medicine practitioners, a recurring theme features that two systems of knowledge are incommensurable, which by definition means that medical and pharmaceutical principles and practices in Korean medicine cannot be translated into scientific and biomedical terminology and principles. Thus, for Korean medicine practitioners, however it is transformed, herbal medicines belong to the professional territory of Korean medicine doctors. In contrast, biomedicine practitioners treat scientific methods used as a key to define the identity of commercialized herbal products (Ma 2010).

I argue that reverse pharmacology has been proposed and promoted, especially by the government, as a practical means to reconcile the alleged incommensurability between the two medical systems and the incommunicability between the two professional organizations of medicines. Reverse pharmacology refers to processes and methods to reformulate and simplify Korean medicinal compositions in order to create new "traditional" pharmaceutical or health products for the biomedical disorders of an international, as well as Korean clientele.[2] Reverse pharmacology is a strategic approach that makes use of traditional medical and pharmacological knowledge. Based on time-tested herbal formulas and clinical experiential knowledge, it identifies candidate plants and pharmacologically active components to develop new drugs. By financially supporting research and academic institutions that aggressively employ reverse pharmacological approaches to herbal medicines, the South Korean government has adopted medical and pharmaceutical bilingualism as an official policy guideline. Accordingly, in the domain of public policy, science or scientific method(s) serves as an instrument to set up regulatory practices and guidelines. Situated in these political and social contexts, in the following I ask a series of questions with a focus on public outbreak over functional health products. Health functional food is legally defined as a product intended to enhance and preserve the human health with one or more functional ingredients or constituents. Health functional foods are very much akin to functional food or functional health products in international contexts (Menrad 2003; Siró et al. 2008), in that they are consumed to prevent nutrition-related diseases and increase physical and mental well-being.

How is safety perceived and practiced when dealing with natural products with health claims? How is safety managed across diverse social groups? More to the point, can scientific methods be employed as a safeguard to ascertain safety and efficacy of health products? In the same vein, does a regulatory agency serve as a gatekeeper to keep adulterated foods from entering into the consumer market? More fundamentally, does the regulatory body have capacity to employ sound science either to falsify or corroborate alleged scientific claims made by multiple parities with conflicts of interest? This chapter takes up these questions with focus on a public outbreak over the safety and efficacy of health functional foods in South Korea, which took place in April 2015. Based on documentary analysis of the mass media, professional articles, patent claims, and governmental and regulatory dossiers, this paper unfolds a complex, interconnected web of market, regulation, technology, venture businesses, government, and consumers in South Korea.

The *Baekshuoh* disaster

On 22 April 2015, Korea Consumer Agency (KCA) publicized that about 90% of functional foods containing *Cynanchum wilfordii* (*Baekshuoh*) on the market were not made out of "authentic" plants (Korea Consumer Agency 2015). The Agency's disclosure was translated into general public as indicating that *Baekshuoh* products on sale were "not safe" to take in. Immediately after the press release, the public was at a loss wondering about what to do with *Baekshuoh* products they purchased and consumed. A matter of fact at heart was rather simple and straightforward: was it safe to consume not-100%-*Baekshuoh* products? Concerned public looked at regulatory authorities for answers. In particular, Ministry of Food and Drug Safety (MFDS) was at the center, being responsible to communicate with the public about safety, toxicity, and any sort of public health-related matters. However, the Ministry failed to communicate with the concerned public by not being able to provide concrete answers to such questions. Accordingly, the Ministry was under public scrutiny regarding its regulatory standards and processes. In particular, a public hearing was convened to investigate the *Baekshuoh* disaster at the Korea National Assembly (The National Assembly of the Republic of Korea 2015). While the media coverage competitively exposed the uncertainty of functional foods, the public hearing turned into a salient stage where regulatory practices were made public.

The KCA is a government organization established in July 1987 in accordance with the Consumer Protection Act, which can be understood as part of reaction to increased public voice for democratic governance. The Agency was expected to serve as a bridge between top-down authoritarian policymaking processes and civic participation. As a governmental agency, its role and function is to protect consumer rights and interests rather than representing governmental and industrial perspectives (Joo 2017; KCA n.d.).[3] Since its foundation, the Agency has remained unnoticed and would have remained invisible without the *Baekshuoh* disaster (Kim and Lee 2015). For years, functional foods targeted at aging women have overflowed into the market and been widely consumed. As a watchdog agency to protect consumers' interests, the KCA looked into the increased production and consumption of women's functional foods that claimed to alleviate menopause symptoms and enhance immune system, and anti-aging. What prompted the KCA's investigation into *Baekshuoh* products was alarmingly a great deal of consumer complaints associated with *Baekshuoh* products. Out of about 1,700 complaints, around 17% were related to *Baekshuoh* products, their side effects, or their adverse effects (Korea Consumer Agency 2015). The KCA collaborated with two prosecutorial and police offices, Seoul Western District Prosecutor's Office and Special Judicial Police Unit of Gyeonggi Province. In concert, they sampled 32 different *Baekshuoh* products that were widely consumed and conducted DNA analysis to determine whether they contained authentic *Baekshuoh*. To their surprise, only about 9% of the sampled products (three out of 32 products) have proved to contain authentic *Baekshuoh*, and the others either contained "*Iyeobupiso*" (similar to *Baekshuoh* in appearance) or an element hardly determined as *Baekshuoh*. The KCA made a particular note about the nature of *Iyeobupiso* in comparison with *Baekshuoh* in its origin and functionality. *Baekshuoh*, also known as *Eunjorong* in Korea, is derived from the roots of *Cynanchum wilfordii Hemsley*, while *Iyeobupiso* is the roots of *Cynanchum auriculatum Royle ex Wight*. Both plants look very similar, even to the trained eye, but *Iyeobupiso* is not listed in the *Korea Herbal Pharmacopeia* for use as a functional food.

More notably, the KCA warned against the use of *Iyeobupiso* as a raw material for foods for the following two reasons. First, the Agency noted that a research suggested that *Iyeobupiso* might cause unpredictable side effects such as hepatotoxicity, nervous breakdown, and weight loss. Second, in the national context of South Korea, there is no evidence established to suggest that *Iyeobupiso* is safe to use for food materials (Korea Consumer Agency 2015). As soon as the Agency's news was released, consumers were panicked, and stock prices of a star venture business, Naturalendo Tech company (NeT), plummeted. MFDS immediately responded to the Agency's claim, leading independent investigation into the same *Baekshuoh* products on sale to concur with the Agency's findings. And yet MFDS was not in sync with the KCA over the interpretation about the safety and potential hazards of adulterated *Baekshuoh* products. Weeks later, on 6 May, NeT released a public apology admitting that the company's product contained "*Iyeobupiso*." As it turned out that "non-authentic" raw materials were used,[4] the company pledged to keep airtight control of quality and suggested implementing new methods to secure quality of raw and processed materials.

Regulatory agency and functional foods

In South Korea, dietary and nutritional supplements were under regulatory scrutiny in accordance with Food Sanitation Act that was enacted in 1962. As the consumer market for dietary and nutritional supplements expanded (Han 1997), there was a growing concern over the safety of health-related products. It was reported that consumers tended to take health claims seriously that were put on labels, partly because of exaggerated advertisement. Legislators lobbied the government to establish the standards and specifications to manufacture and distribute products with health claims.[5]

In 2002, the Health Functional Food Act was enacted to ensure the safety and functionality of health functional foods with certain health claims for consumer information. A new category, "health functional food," was introduced to establish a distinction between food and drug under Health Functional Foods Act, and put into implementation about a year later, in 2003. The law requires manufacturers to provide scientifically valid evidence to corroborate claimed health benefits in one of the following four categories: nutritional function, improvement of health, and reduced susceptibility to diseases (or enhanced immune system) (ACT 2011).[6]

In 2013, when the incumbent presidential administration took office, it defined adulterated food as one of four axes of evil in society to be eliminated. To that cause, the government promoted Korea Food and Drug Administration to the MFDS, granting more administrative authority to oversee and control safety issues over food and drugs. That administrative action was the expression of the government's determined will to keep tight control of food quality and safety. MFDS used to be an agency under the supervision of the Ministry of Health & Welfare (MOHW). Now MFDS exists and functions independently of MOHW. What this means is that MFDS holds sole responsibilities to control and govern safety networks of foods and drugs with utmost power.

When the *Baekshuoh* disaster came to the surface, the emblematic symbols of international and domestic recognition became under fire, as if opening Pandora's Box. The question at stake was rather simple: do *Baekshuoh* products contain *Baekshuoh*? Ironically, this naïvely straightforward question has upended black-boxed processes of regulating functional foods and drugs. Consumers who had been consuming *Baekshuoh* products were anxious to know whether they had been taking in adulterated foods. The media and the public looked to the MFDS for an answer, with expectations that the Ministry would be able to confirm: first, whether or not *Baekshuoh* products on sale were adulterated with the mix of not-officially-recognized herb (that is, *Iyeobupiso*); and second, whether the consumption of adulterated foods was, in scientific terms, harmful to the body and if so, in what way?

A biotechnology company and natural health products

On 26 April 2010, Korea Food and Drug Administration (KFDA) certified a complex herbal compound containing extracts of *Cynanchum wilfordii*, together with *Phlomis umbrosa* and *Angelica gigas*, as a functional ingredient for health and functional food in compliance with Articles 14 and 15 of the Health Functional Food Act of Republic of Korea. A functional ingredient refers to a substance providing health benefits, such as a processed raw material, extract, purified substance, or combination of the former ingredients from animal, plant, or microorganism. The compound is known as EstroG-100 and is produced by the biotechnology company NeT.

After its first failed attempt in 2008, NeT sought approval again from the KFDA, claiming that EstroG-100 would enhance women's health with climacteric symptoms (Naturalendo Tech 2010). Since its foundation in 2001, the company has primarily focused on the development of endocrine commodities by transforming plant resources into commercial products such as phytoestrogen. NeT's rather smooth entry into the market (or rise as a star venture biotechnology company) can be attributed to the South Korean government's ambitious goal to emerge as a competitive player in biotechnology fields in the global market. From the very start, the company collaborated with the university research team under government sponsorship to research and develop functional elements effective to help with menopausal symptoms. In place of synthesized remedies, they turned into natural plants, making use of traditional medical and pharmaceutical knowledge. Its first market debut was made with Estromon®, a claimed natural alternative to pre-existing hormone replacement therapy (HRT) used to alleviate the menopausal symptoms. HRT is known to be effective for the climacteric symptoms such as insomnia, hot flushes, night sweats, memory loss, and depression. It is also well established that HRT prevents reproductive organ disorders such as urinary incontinence and vaginal dryness, and reduces the risks of osteoporosis and bone fracture (Lee et al. 2005).

Estrogen is a female hormone which regulates the menstrual cycle and the reproductive system, and the growth and development of female secondary sexual characteristics such as breasts and endometrium. As women enter into a period of menopause, estrogen deficiency is visible with age and presents such symptoms as hot flushes derived from vasomotor instability, urogenital degeneration, osteoporosis, arteriosclerosis, and coronary heart disease, and the like. Because the patterns of estrogen secretion are directly related to the advancement of aging, women around the stage of menopause suffer from various kinds of symptoms (Kim 2006, 2007, 2010). As Women's Health Initiative (WHI) began to look into the effectiveness and safety of hormone therapy at the global level, there was also reaction from South Korean researchers who began to look for natural therapies. In 2003, local medicinal plants researchers took interest in *Baekshuoh* to see whether it is effective and safe to treat or ameliorate symptoms associated with menopause. Through a series of patent claims and clinical trials, NeT articulated a range of inventions of EstroG-100 (Kim 2006, 2007, 2010; Chang et al. 2012). The patents are composed of two major claims, a new composition of herbal compounds and a method to prepare it. EstroG-100 is comprised of the three botanical extracts and helps to treat, prevent, and ameliorate symptoms associated with menopause. The three extracts are mixed in almost equal proportion by weight (*Cynanchum wilfordii* 32.5%, *Phlomis umbrosa* 32.5%, *Angelica gigas* 35%). Each of the three botanical extracts has key phytochemical characteristics that are unique to each plant (Kim 2010; US Food and Drug Administration 2011).

To translate the findings into marketable products, Korea Institute of Planning and Evaluation for Technology in Food, Agriculture, Forestry, and Fisheries funded the research and development project titled "Globalization of proprietary standardised phytoestrogen by the study of active ingredients and clinical effectiveness on Caucasians" as part of development projects in agriculture and forestry technologies. Between 2008 and 2010, 374 million KRW had been poured into the project to extract and crystallize an efficacious ingredient from a complex mixture of three medicinal plants. Eventually, the research team of NeT discovered that EstroG-100 improved 10–12 typical menopause-associated symptoms such as hot flushes, vaginal dryness, paresthesia, insomnia, nervousness, vertigo, melancholia, fatigue, rheumatic pain, and formication. According to the research team, clinical studies confirmed that EstroG-100 was safe without major side effects such as changes in serum estrogen, follicle stimulating hormone, body weight, and bone density were not reported (iPET 2012).

In 2009, the South Korean government awarded the Jang Young-Shil Prize to the research team of NeT in recognition of inventiveness of the technology and the product (iPET 2012). Jang Young-Shil is known to be the greatest inventor in Korean history, and in 1991 the government made this prize named to encourage engineers' inventive activities and the commercialization of new inventions (IR52 n.d.).[7] EstroG-100 was awarded as one of Korea's ten best new technologies of 2013, being recognized both with the potential for market growth and with innovative creativity and therapeutic potential. Not surprisingly, most awardees of that year were identified with giant conglomerates such as Samsung Electronics, SK Chemical, and LG Electronics. Every year, the government selects the ten best new technologies that are original, innovative, and marketable, and EstroG-100 was understood to represent the most successful case: it was developed by a biotechnology venture company (as opposed to big conglomerates), and it was commercialized to be sold on global markets. The incumbent government has emphasized creative economy, for which the development and commercialization of innovative technologies is taken to play a critical role. Thus, in 2014, the government nominated and awarded NeT as an exemplary company in leading creative economy (Naturalendo Tech 2015). Besides, Naturalendo Tech is decorated with many more national awards, which also served as a justification to guarantee the safety and effectiveness of EstroG-100.

The company also targeted global markets. In March 2006, NeT submitted to the US Food and Drug Administration (FDA) the notification of a new dietary supplement pursuant to 21 U.S.C. 350b(a)(2) (section 413 [a][2] of the Federal Food, Drug, and Cosmetic Act). FDA is expected to act as a gatekeeper to keep adulterated dietary supplements from being ingested by consumers. The US federal act requires that either manufacturer or distributor of a dietary supplement submit evidential information necessary to determine whether or not the claimed substance is reasonably safe to consume. The manufacturer is responsible for providing the evidence of safety that a dietary supplement containing a new ingredient has the claimed health effects, at least 75 days before interstate commerce. After having reviewed the submitted documents from the US partner of NeT, FDA concluded that the submitted information was not sufficient to determine the identity of EstroG-100, on the grounds that the company failed to demonstrate "which parts of *Cynanchum wilfordii* or *Phlomis umbrosa* are used and how the roots of *Angelica gigas* and the other two plants are processed or extracted to produce EstroG-100" (US Food and Drug Administration 2006). However, with its second trial in 2010, USFDA approved EstroG-100 as new dietary ingredient (US Food and Drug Administration 2011), and in the following year in July 2011, Health Canada granted a Natural Product Number (Naturalendo Tech 2013).

Once the regulatory barrier at domestic and international levels was cleared, the company diversified distribution channels and entered into the markets through giant franchised supermarkets such as Walgreens, Rite Aid, and Walmart, and direct-to-consumer sales via cable television channels. As a marketing strategy, NeT used international and domestic regulatory agencies' recognition as corroborating evidence that its product is safe and effective to consume, on the one hand. The fact that FDA "permitted" EstroG-100 containing products to market in the United States was employed to imply to the Korean public that FDA "approved" the safety and effectiveness of *Baekshuoh*. With the opening of the North American markets, the company aggressively knocked on Asian and European markets (Naturalendo Tech 2015). At the domestic market, to pitch the sales of its product the company advertised that its product won many prizes awarded by the South Korean government.[8] On the other hand, NeT invoked in the public's mind that its product is based on historical records, i.e., *Donguibogam*, when aired via documentary and shopping channels.[9] According to the NeT's report, the domestic consumer market for female hormones have expanded dramatically from about 70 billion KRW in 2010 to about 310 billion KRW in 2013 thanks to EstroG-100 (Naturalendo Tech 2013). The market success of NeT's technology and products has contributed both to increasing farm households' income and to vitalizing biotechnology venture companies (iPET 2012).

Baekshuoh (白首烏, *Cynanchum wilfordii*)

What is *Baekshuoh*? *Baekshuoh* is also known as *Eunjorong*, *Cynanchum wilfordii*, and, by extension, as EstroG-100, depending on the context of circulation. When the medicinal plant is brought into contact with various social groups through circulation, its identity changes. How is it authenticated in light of safety, quality, and efficacy? How does it circulate within and beyond the domestic market? What is it that governs circulation?

First, *Baekshuoh*, a natural plant, has been translated into a trade name EstroG-100 (or EstroG) via contemporary science. As described earlier in this chapter, EstroG-100 is a herbal product containing a mixture of standardized extracts of *Phlomis umbrosa* (韓續斷), *Cynanchum wilfordii* (白首烏), and *Angelica gigas* (當歸). The targeted consumer groups are those women suffering from peri-, pre-, and post-menopausal symptoms, as all three medicinal plants, respectively, have been used to improve health. According to the information offered by the company, randomized, double-blind clinical trials were conducted both at one of major hospitals in Seoul and abroad, whose results supported the functionality of EstroG-100 without major side effects (Chang et al. 2012; Lee et al. 2005). The company submitted the evidence of safety, adducing the results of animal toxicity tests and clinical studies of human beings conducted at domestic and international hospitals. In the United States, EstroG-100 is sold via online and offline marketing channels such as Walgreens, CVS Pharmacy, Whole Foods, Sprouts, amazon, and iHerb. As advertised on the website of the marketer, EstroG-100 is sold as a lifestyle nutritional supplement.

> For so long women have feared menopause and suffered through it just to feel like they're never quite the same again – like they've lost their normal lives. All that's been available to treat the symptoms are herbal remedies that work at a snail's pace or not at all. With EstroG-100, we finally have a solution that works, and the speed with which it works is unheard of.
>
> (Jeffers 2015)

Equally importantly, historical evidence was also adduced to demonstrate that the medicinal plant *Cynanchum wilfordii* had been applied to the human bodies for therapeutic purposes. Traditional medicine texts are referenced to authenticate therapeutic properties and clinical uses of a medicinal herb. As many countries in East Asia share the tradition of Chinese medicine, pharmacopeias from various countries with shared tradition are referenced.

> This particular Korean herbal medicine is also recorded in *Donguibogam* (Heo, 1610) as *Eunjorong*, and its applications included anemia, weakness after disease as a recovery agent, weak muscle and bones and white hair. According to the most famous and oft-cited traditional medical handbook in Korea, it is safe to use from a toxicological point of view. There is an orally transmitted tale that a man who had eaten the dried roots of *C. wilfordii* gained libido 7 days later and recovered from various diseases after 100 days. Ever since, oriental medicine doctors have prescribed it widely in Korea. Now, it is one of most frequently prescribed herbal medicines in Korean oriental hospitals
>
> (US Food and Drug Administration 2006).

In South Korea, *Donguibogam* is valuated as the most authoritative classical medicine text in existence, both to the expert community and to general public. In a way, the text has been reincarnated as a modern legend to the extent that the text is taken to be authoritative, and whose factual status is rarely questioned (Lee and Baek 2014; Kim et al. 2006). As much as *Donguibogam* is referenced for therapeutic and diagnostic purposes in the Korean medicine community, the classical herbal formulas have also been studied to apply for dietary and nutritional purposes (Kim 2014). Referring to the history of use of the herbs that are recorded in classical medicine texts, NeT appealed to culturally deep-rooted faith in Korean medicines.

As technological artifacts or ideas are displaced from the original site and in circulation, there occurs transformation, adaptation, reinterpretation, and reutilization in accordance with local environments. Looking in history, oftentimes imperial governments had imposed on the indigenous societies the adoption of new inventions, whether legal, societal, or technological, which reversely offered local communities with culture-specific rationales to react against such universalizing forces. In the Northeast Asian region, Chinese medicine is widely shared and practiced with local variations. In South Korea, the nationalization of traditional medicine as Korean medicine has progressed since independence from Japanese colonial rule in 1945. Practitioners of Korean medicine have made strategic efforts to systematize their practices in order to stand rather independently from Chinese medicine, Japanese *Kampo* medicine, and other Asian medical systems. Hence, whenever a question arises over the authenticity of Korean medicines derived from medicinal plants, there is a classificatory correspondence issue at stake if the raw plant and its therapeutic uses is not recorded in the official pharmacopeia. *The Korean Herbal Pharmacopeia* is an officially endorsed pharmacopeia that registers all medicines, with references to classical Korean medicine texts. The *Pharmacopeia* records the herbs that are used as medicines singularly or in combination with other herbs. That is, a medicinal plant should be recorded in the *Pharmacopeia* to be used as a raw material for food or pharmaceuticals. The *Pharmacopeia* provides information about the herb, ranging from the origin, botanical nomenclature, and physiognomic qualities such as shape, size, color, and taste, to microscopic qualities and compositional information at the cellular level. The *Pharmacopeia* also contains information about the differential weight between raw material and dried material, the identification method, and how to extract efficacious elements from herbs. The herbs in the *Pharmacopeia* are prepared and certified in accordance with standards manufacturing processes to assure and control production and quality management. In cases of herbs, manufacturing processes are applied to selecting herbs, drying, cutting, inspecting, and packaging.

As to the medicinal plant under discussion, the *Pharmacopeia* records that *Baekshuoh* is a tuberous root of *Eunjorong* (*Cynanchum wilfordii*), which is native to the Korean peninsula. However, it is known that the plant is neither registered nor recognized in Chinese, Taiwanese, and Japanese pharmacopeias under the same name (Doh et al. 2015). The name of *Baekshuoh* is present in the Chinese medicines text 山東中藥 (1959), although it is uncertain that the herb refers to the same plant as present in Korean medicines texts (Lee and Kweon 2012). Though the plant is registered as *Baekshuoh* in the *Pharmacopeia*, it is circulated on the market under three different names: *Baekshuoh* (白首烏), *Hashuoh* (何首烏), and *Baekhashuoh* (白何首烏). These appellations are indiscriminately used among herb retailers and cultivators. As the market demand for *Baekshuoh* increased in the 1990s because of the market expansion of health-related products, the retail market turned into the managed cultivation of the herb in the 1990s. About that time, *Iyeopupiso* was imported from China, and there is no historical record in existence that the herb was used for therapeutic purposes in Korean medicine. According to the tradition of Korean medicine (東醫寶鑑, 東醫壽世保元), *Baekshuoh* has been used for the purposes of enhancing strength, boosting energy, and nourishing the blood.

Iyeopupiso is said to be morphologically very similar to *Baekshuoh*, though its precise therapeutic functions and medical efficacy are not written down in medicines texts. As compared with *Baekshuoh*, *Iyeopupiso* produces a higher volume of roots, and is thereby preferred for cultivation by farmers for economic reasons. Farmers began to cultivate it instead of *Baekshuoh* in the 1990s. Accordingly, in 2010, the Korea Food and Drug Administration intervened to set up identification methods so as to distinguish the two plants even before the public outbreak took place. For instance, *Baekshuoh* – in contrast with *Iyeopupiso* (*Cynanchi auriculati*) – contains a component called "conduritol F," and its presence and concentration can be detected via scientific methods such as thin film chromatography (TLC) or high-performance liquid chromatography (HPLC). The micro characteristics of the herb were subsequently written into the *Korean Herbal Pharmacopeia* in 2011 (Doh et al. 2015; Kim et al. 2015; Lee and Kweon 2012). In addition to the analysis of medicinal plants, cultivation methods were also systematically studied to differentiate *Baekshuoh* from contending herbs (Kim et al. 2014).

Until the public disaster took place, scientific research was available to suggest that *Iyeopupiso* was different from *Baekshuoh* at the cellular level, information which has not yet translated into regulatory considerations. The two plants were circulated from cultivation to sales, which brought a chaotic situation when their food products needed scientific corroboration.

Conclusion: circulation and governance of Asian medicine

The case under discussion should be understood in the context of government-driven winning market strategies, which became visible and apparent during the early 1990s, particularly in terms of allocation of governmental resources. The government financially and institutionally supported research on herbal medicines to turn them into marketable products, whether patented medicines or evidence-based health products. At one level, the government restructured legal and regulatory infrastructures to facilitate the commercialization and pharmaceuticalization of Korean medicines. At another level, the government sponsored collaborative research among the pharmaceutical industry, the academy and research institutes to promote the translation of Korean/herbal medicines into marketable pharmaceuticals. The government's policy initiative to "modernize" herbal medicines was based on the understanding that science or science-based medicines would prevail in the international markets. In particular, the government supported research and academic institutions that aggressively employed reverse pharmacological approaches that reutilized and repackaged knowledge of Korean medicine. The South Korean government has adopted medical and pharmaceutical bilingualism as an official policy guideline. Although governmental actions ignited deep-rooted conflicts between the supporters of two medical/pharmaceutical systems, i.e., Korean medicine and Western biomedicine, scientific methods serve as an instrument to set up regulatory practices and guidelines for public policy. As the pharmaceutical industry and biotechnology companies, under the aegis of the government, jumped into "the reformulation regime" (Pordié and Gaudillière 2014, p. 63) for Korean medicines, they allied with medical and pharmaceutical communities trained in biomedical sciences and came up with new manufacturing and technical tools and methods to place their inventions in a biomedically defined social world.

The intersections between science and democracy have long received scholarly attention, highlighting the politicized aspects of techno-science in various social and historical contexts (Bowker and Star 1999; Epstein 1996; Ezrahi 1990; Hess 2007; Jasanoff 1998, 2007; Merton 1973 [1942]). Challenging a common myth that science is a value-neutral enterprise that can be employed as a basis for organizing the social and political order, this body of scholarship eloquently demonstrates the contextualized co-production of political and scientific orders in a given society. What is particularly germane to the case under discussion is the intersection between science and policy, dealing with criteria that serve as bases for regulatory mechanisms. Science and scientific methods are employed to set up regulatory and legal guidelines. Regulatory science is subject to a loose and fluid definition, and it is accountable to non-scientific watchdogs such as the media, interested public, and the government (Jasanoff 1998). The guidelines for validating science in the regulatory context tend to be fluid, controversial, and arguably more politically motivated than those applicable to university-based research (Irwin et al. 1997; Jasanoff 1998).

As much as regulatory agencies have emerged as a locus for scientific fact-finding and for adjudicating controversies about science in Europe and the United States, the South Korean government has followed suit. In particular, it has realigned the administrative structure to reinforce the role and function of regulatory agencies in the fields of public health and foods. When the *Baekshuoh* controversy took place in April 2015, the very nature of regulatory agency and practices was under public scrutiny. Regulatory agencies, by definition, are expected to assess and evaluate the status of scientific facts and to determine the safety of the product in question, let alone quality control and supervision of circulation and sales. To a certain degree, the regulatory agency has remained black-boxed until the *Baekshuoh* disaster took place. When the agency could not provide clear-cut answers and not make advisory decisions for the best interests of the public on the basis of scientific evidence, the public – as well as the government – accidentally became involved in de-constructing and reconstructing the meanings of safety in regulatory practices. Before the *Baekshuoh* disaster, safety was a given on the basis of regulatory agencies' recognition as the information about the safety flew from top-down. And the mass media (TVs, newspapers, commercial advertisements) reproduced and reified the safe image of *Baekshuoh*. The *Baekshuoh* disaster opened widely the very presumed/accepted processes of establishing safety and effectiveness of health foods in South Korea.

Notes

1 In fact, it is a controversial matter whether the team leader should get credited most.
2 On this method and its larger implications, see Pordié and Gaudillière (2014).
3 http://english.kca.go.kr/wpge/m_24/en4210.do

4 No concrete evidence in literature has been established to prove that *Iyeobupiso* is harmful to humans. KCA and MFDS maintained a conflicting interpretation over the safety or hazards of *Iyeobupiso*. Both agencies agreed that it is not advisable to use *Iyeobupiso* as a raw material for human food use because its safety has not been scientifically established yet.
5 Health Functional Food Code. www.mfds.go.kr/files/upload/eng/4.Health_Functioanl_Food_Code_(2010.09).pdf; www.mfds.go.kr/eng/index.do?nMenuCode=65
6 www.mfds.go.kr/eng/index.do?nMenuCode=65
7 www.ir52.com/eng/ir52award.asp
8 Anecdotally, there is an indicator how far the *Baekshuoh* disaster reached out to the Korean society in a relatively short period of time. I had a chance to have a casual conversation with colleagues from work about the *Baekshuoh* disaster. A senior colleague of mine mentioned to me that his wife had bought *Baekshuoh* products via home shopping cable channels and been taking them for a long time. When I asked if had felt any difference with *Baekshuoh*, he was certain that his wife had experienced the difference and felt the need to continue taking it in. However, soon after the KCA's press release, she threw out all remaining products. Following his story, another colleague stated that her father was greatly concerned because the stocks of NeT he had bought kept going down.
9 www.naturalendo.co.kr

References

Bowker, Geoffrey C., and Susan Leigh Star. 1999. *Sorting Things Out: Classification and Its Consequences.* Cambridge, MA: The MIT Press.

Callaway, Ewen, and David Cyranoski. 2015. "Anti-Parasite Drugs Sweep Nobel Prize in Medicine 2015." *Nature* 526: 174–175. doi: 10.1038/nature.2015.18507.

Chang, Albert, Bo-Yeon Kwak, Kwontaek Yi, and Jae-Soo Kim. 2012. "The Effect of Herbal Extract (EstroG-100) on Pre-, Peri- and Post-Menoapusal Women: A Randomized Double-Blind Placebo-Controlled Study." *Phytotherapy Research* 26: 510–516.

Doh, Eui-Jeong, et al. 2015. "Microscopic Identification-Keys for Cynanchi Wilfordii Radix and Cynanchi Auriculati Radix." *Korean Journal of Herbology* 30 (4): 65–69.

Epstein, Steven. 1996. *Impure Science: AIDS, Activism, and the Politics of Knowledge.* Berkeley, Los Angeles, and London: University of California Press.

Ezrahi, Yaron. 1990. *The Sescent of Icarus: Science and the Transformation of Contemporary Democracy.* Cambridge, MA: Harvard University Press.

Han, Gil Soo. 1997. "The Rise of Western Medicine and Revival of Traditional Medicine in Korea: A Brief History." *Korean Studies* 21: 96–121.

Hanson, Marta. 2015. "Is the 2015 Nobel Prize a Turning Point for Traditional Chinese Medicine?" *The Conversation*, October 6. https://theconversation.com/is-the-2015-nobel-prize-a-turning-point-for-traditional-chinese-medicine-48643.

Hess, David. 2007. *Alternative Pathways in Science and Industry: Activism, Innovation, and the Environment in an Era of Globalization.* Cambridge, MA: The MIT Press.

INS (Institute of Natural Materials, Seoul National University). 1998. *Multidisciplinary New Drugs Discovery System from Natural Products* (Traditional Korean Medicinal Materials). Seoul: Ministry of Science and Technology.

iPET (Korea Institute of Planning and Evaluation for Technology in Food, Agriculture, Forestry & Fisheries). 2012. *Development and Globalization of Health Functional Food Ingredient for Menopausal Symptoms.* January 2. https://eng.ipet.re.kr/News/NewsVP.asp?tbl_id=.

IR52. N.d. "IR52 Jang Young-Shil Award?," https://www.ir52.com/eng/ir52award.asp. Accessed July 05, 2019.

Irwin, Alan, Henry Rothstein, Steven Yearley, and Elaine McCarthy. 1997. "Regulatory Science: Toward a Sociological Framework." *Futures* 29 (1): 17–31.

Jang, Il Moo. 2001. *Development of Japanese Version of Traditional Oriental Medicines Database*. Seoul: Ministry of Health and Welfare.

Jang, Il Moo. 2011. "Cheonyeonmul Sinyak: Mwon Yeakinga?" (Natural Drugs: What Drugs AreThey?). *Bioin* 20: 1–17.

Jasanoff, Sheila. 1998. *The Fifth Branch: Science Advisers as Policymakers*. Cambridge, MA: Harvard University Press.

Jasanoff, Sheila. 2007. *Designs on Nature: Science and Democracy in Europe and the United States*. Princeton, NJ: Princeton University Press.

Jeffers, Michael G. 2015. *EstroG 100*. September 8. estrog100.com/what_is_estrog100. Accessed September 8, 2015.

Joo, Sungsoo. 2017. "Citizen Participation and Democracy in the History of Korean Civil Society (1987–2017)." *Civil Society & NGO* 15(1): 5–38.

KCA (Korea Consumer Agency). "Purpose of Establishment." N.d. http://english.kca. go.kr/wpge/m_24/en4210.do. Accessed July 04, 2019.

KCA (Korea Consumer Agency). 2015. *Most Baekshuoh Products in Market Circulation Are Fakes*. Press release, Seoul: Korea Consumer Agency.

KDRA (Korea Drug Research Association). 1992. *Cheonyeonmul Robuteoui Sinyak Gaebal* (Studyon the Development of New Drugs from Natural Products). Seoul: Ministry of Science and Technology.

Kim, In-Jae, et al. 2014. "Seasonal Change of Growth Asclepiadaceae Plants (Cynanchum wilfordii, C. auriculatum, Metaplexis japonica)." *The Korean Society of International Agriculture* 26 (3): 292–296.

Kim, Jae-Soo. 2006. Method for Treating or Preventing Symptoms Associated with Menopause. United States Patent US20060193929 A1, filed August 31.

Kim, Jae-Soo. 2007. Method for Accelerating Secretion of Estrogen and Regenerating Tissue Cells of Female Sexual Organs. United States Patent US20070269538 A1, filed November 22.

Kim, Jae-Soo. 2010. Method for Treating or Preventing Symptoms Associated with Menopause. United States Patent UD 7763284 B2, granted July 27.

Kim, Jong Du. 2014. "A Study in Korea's Diet and Preservation Health of Donguibogam." *Korean Culture and Thought* 72: 337–355.

Kim, Kyu-Heon, et al. 2015. "Development of Primer Sets for the Detection of Polygonum Multiflorum, Cynanchum Wilfordii and C. Auriculatum." *Journal of Food Hygiene and Safety* 30 (3): 289–294.

Kim, Nakhyung, et al. 2006. "A Critical Review on the Communication Disorders in Donguibogam." *Communication Sciences and Disorders* 11 (3): 113–128.

Kim, Yong Sik, and Lee Seong Taek. 2015. "Home Shopping Channels Who Exaggerated Advertisement about *Baekshuoh* Should Fully Refund Consumers." *Hankook Ilbo*, May 18.

ACT. 2011. "Health Functional Foods Act." http://www.moleg.go.kr/english/korLawEng? pstSeq=58332&pageIndex=6. Accessed July 04, 2019.

Lee, Dong-Woo, and Jin-Ung Baek. 2014. "A Study on Analyzing the Terms Describing Anti-Aging Effects in Dongeuibogam to Propose the Methodology for Selecting Medicinal Herbs Related to Anti-Aging Effects." *Journal of Oriental Medical Classics* 27 (2): 25–48. doi: 10.14369/skmc.2014.27.2.025.

Lee, Je-Hyun, and Kee-Tae Kweon. 2012. "Determination of Harvest Time and Nominal Origin from Cynanchi Wilfordii Radix." *The Journal of Korean Oriental Medicine* 33 (1): 160–168.

Lee, Ki Ho, Duck Joo Lee, Sang Man Kim, Sang Hyeun Je, Eun Ki Kim, Hae Seung Han, and In Kwon Han. 2005. "Evaluation of Effectiveness and Safety of Natural Plants Extract (Estromon) on Perimenopausal Women for 1 Year." *Korean Society of Menopause* 11 (1): 116–126.

Ma, Eunjeong. 2010. "The Medicine Cabinet: Korean Medicine under Dispute." *East Asian Science, Technology, and Society* 4 (3): 367–382. doi: 10.1007/s12280-010-9147-9.

Ma, Eunjeong. 2015. "Join or Be Excluded from Biomedicine? JOINS® and Post-Colonial Korea." *Anthropology & Medicine* 22 (1): 64–74. doi: 10.1080/13648470.2015.1004774.

Menrad, Klaus. 2003. "Market and Marketing of Functional Food in Europe." *Journal of Food Engineering* 56: 181–188. doi: 10.1016/S0260-8774(02)00247-9.

Merton, Robert K. 1973 [1942]. "The Normative Structure of Science." Chapter 3 In *The Sociology of Science: Theoretical and Empirical Investigations*. Chicago: University of Chicago Press.

The National Assembly of the Republic of Korea. 2015. "Adhoc Minutes of Ministry of Health & Welfare." *Pending Issues on Raw Materials of Baekshuoh Products*. Seoul, May 6.

Naturalendo Tech. 2015. "EstroG, Hot on Canadian Public News Media CBC," January 26. www.naturalendo.co.kr.

Naturalendo Tech. 2010. Certificate of Functional Ingredient for Health/Functional Food. South Korea Patent KFDA FID 000000-00001, granted April 26.

Naturalendo Tech. 2013. *New Proprietary Standardized Phytoestrogen*. Electronic document, Pangyo and Korea: Naturalendo Tech Co., Ltd.

Pordié, Laurent, and Jean-Paul Gaudillière. 2014. "The Reformulation Regime in Drug Industry: Revisiting Polyherbals and Property Rights in the Ayurvedic Industry." *East Asian Science, Technology and Society* 8 (1): 57–79.

Siró, Istvá, Emese Kápolna, Beáta Kápolna, and Andrea Lugasi. 2008. "Functional Food. Product Development, Marketing and Consumer Acceptance: A Review." *Appetite* 51: 456–467.

US Food and Drug Administration. 2006. *Memorandum: 75-Day Premarket Notification of New Dietary Ingredients: Memorandum*. College Park, MD: Public Health Service Food and Drug Administration.

US Food and Drug Administration. 2011. *Memorandum: 75-Day Premarket Notification of New Dietary Ingredients: Memorandum*. College Park, MD: US Food and Drug Administration.

3 Governing medical traditions in Myanmar

Céline Coderey

Products from local traditional pharmacopeia are ubiquitous in Myanmar.[1] There is almost no family without a stock of traditional products for the prevention or cure of common or chronic ailments, and when people from Myanmar move abroad, one of the first things on their agenda is to ensure a supply of the remedies they are so familiar with. Composed from ingredients of vegetal, animal, and mineral origin, traditional medicines are available in the form of raw materials, powders, pills, balms, or liquids. Practitioners and market vendors distribute them in little plastic bags or bottles, while pharmacies and private clinics sell ready-made industrial products in eye-catching colorful boxes illustrated with company logos, Kachin mountainous landscapes, royal or Buddhist symbols, or contemporarily outfitted youth.

In this chapter, I examine the diversity of traditional medical products and their wide circulation vis-à-vis the governance system they are inscribed in, and reflect on what these products and their regulations mean in the sociopolitical context of contemporary Myanmar. I view governance in a Foucauldian sense (1991, 1995), as the coexistence and negotiation of different regulations by actors located in different spaces and levels, micro and macro, local and global, and operating according to different logics – political, economic, social, medical, and legal. Governance in this sense includes "all kinds of rules, prescriptions and behaviors, which *in fine* govern social activity." The act of regulation thus also involves "alternative routes, corrupt practices and other forms of arrangements" (Quet et al. 2018, 15).

Traditional medicine has ancient roots in Myanmar, existing since time immemorial as remedies crafted by local healers, using ingredients either self-collected or purchased at the market. The different phases of the "life" of the products (Van der Geest et al. 1996; Reynolds Whyte et al. 2002) were regulated by the medical and commercial concerns of the sole individuals involved. However, post-independence, two major events contributed to expand and intensify the circulation of traditional medicines, while inscribing them in a wider system of governance where a major role is played by the different units of the state – such as the Ministry of Health, the Ministry of Finance, the Ministry of Industry, etc. – and global actors such as the World Health Organization (WHO), the World Trade Organization (WTO), and the Association of Southeast Asian Nations (ASEAN). The first of these events was the development, at a national level, of an institutionalized and standardized version of traditional medicine integrated into the formal national healthcare system, a process initiated by the Burmese[2] government in 1952, in the aftermath of independence from the British (1948), in tune with the nationalistic ethos of that period, with the intention of raising traditional medicine from the abyss into which it had sunk during the colonial period. The second event was the mechanization of the industry and the increasing commercialization of goods concomitant to the expansion of national and international trade which occurred after the 1970s.

It bears underscoring that the expansion of the Myanmar traditional medicine market remains limited when compared to countries like India, China, Tibet, and Vietnam, where the medical market benefited from greater state support, private investors, and a stronger overall economic sector. Sophisticated factories and research labs sprung up, and new and innovative products that throve in the global market were developed. In Myanmar, 50 years of dictatorship (1962–2011) and the accompanying socialist-oriented economy, coupled with the embargoes imposed by many countries, hindered the development of the local economy. Moreover, the state's contribution to the health sector was always minimal; a scant 2.8% of gross domestic product. Traditional medicine received less attention and support than Western biomedicine, greatly limiting any possibility of significant development. If Myanmar traditional medicine does traverse its national borders, it is still solely addressed to the Myanmar diaspora. However, like its Asian counterparts, traditional medicine has been inscribed within the wider system of governance.

Studies that have described this phenomenon for other countries in the region (Wahlberg 2012; Saxer 2013; Pordié 2011) have indicated that the global actors involved in governance support the international homogenization of the regulations to ensure smooth circulation of safe and quality products. The WHO, in particular, acts as the directing and coordinating authority on international health work, providing the individual states with norms and guidelines, primarily drug laws that concern licensing and registration; good manufacturing practices (GMP); good distribution practices (GDP); "border regimes," administrative efforts for regulating the transition of herbs and traders across borders (Saxer 2013, 109); and the licensing of manufacturers, sellers, and practitioners. This homogenization assumes more relevance since it is grounded in standardizing models originally developed in Western countries for the regulation of biomedicine and biomedical pharmaceuticals (Langford 2002; Wahlberg 2012; Saxer 2013; Pordié 2010, 2011; Craig 2012; Adams 2001, 2002; Adams et al. 2011). The "deterritorialization" of these models and their application to other sociocultural contexts and medical epistemologies presupposes the universality and neutrality of biomedical science. This presumed transferability is what makes of them what Collier and Ong (2005) call "global forms." In other words, in order to be legitimized and allowed to circulate in the global market, traditional medicine must be compliant with biomedical standards, attesting to what Foucault (1991, 1995) would call the biopolitics of pharmaceutical governance (Craig 2012, 166). This epistemic hierarchy raises many criticisms, at least from the ethical standpoint (Craig 2012; Saxer 2013). Indeed, the question arises as to what extent traditional medicines should compromise their own principles, identity, and efficacy in order to comply with biomedical criteria (Adams 2001).

While the reach of the state and global actors is an undeniable factor in the regulation of medicines, this does not mean that governance becomes a top-down process, a unidirectional imposition of global forms. The arrival of new actors and regulatory regimes does not lead to the disappearance of the more traditional ones. What occurs is additional layering that overlaps the existing one. Indeed, as Quet et al. (2018) have shown, the different layers coexist and affect one another, converging and diverging on occasion, generating tensions that beg scrutiny. This is all the more true since the plethora of actors – the state, global actors, healers, and manufacturers – do not represent a coherent unit. Within each repose different opinions and values. Neither is the state synchronous with global actors, and nor does it – or healers and manufacturers – passively implement global directions. Indeed, as stressed by Craig (2012), Saxer (2013), and others, GMP and other regulations are not only de-contextualized from the original context, but are "re-contextualized" in new environments and reshaped by local actors in accordance with their own ideas on merit and demerit, cultural values, economic interests, and the material contingencies they live in. Global forms are thus "re-territorialized" in an assemblage (Collier and Ong 2005, 4) producing what Michael Burawoy (2000) calls "grounding globalization," where territory, land, and nations are allowed to express themselves in globalization processes, and play an instrumental role in shaping the global through diversifying and making it more complex.

However, Myanmar conforms to a trajectory less similar. Under authoritarian military dictatorship for almost 50 years, the state has controlled every aspect of the economy, including medicine. Simultaneously, functioning as the sole regulatory authority, it has been a spoke in the wheel of privatization, unlike other countries where private research laboratories, manufacturing companies, and donors played a major role. Governance was determined by the state and the response to it by the healer-manufacturer combine. Moreover, the strong centralization tendency and the domination of the Burmese majority over the ethnic minorities left scant room for the organic development of traditional medicine. Under the circumstances, what does it mean for a totalitarian state to endorse regulations that are intrinsically an instrument of biomedical governance? What is done in the name of biomedical hegemony, and what do these regulations represent for manufacturers and healers?

I will respond by describing the governmental measures to institutionalize and regulate traditional medicine, and illustrate what the biomedicine-based regulations mean for the state. I will also examine the implementation of the project against the backdrop of the political economy of the state. Finally, I will focus on healers and manufacturers and examine their responses to the state's regulations.

Governance: building a national, standardized, modern medicine

Medicine at the service of nation-building

Four years after independence from the British, the Burmese government integrated "traditional medicine" into the formal health system, alongside the colonizers' biomedical model. A Traditional Medical Council was established as the main body responsible for all matters relating to traditional medicine (WHO 2012, 79). The government's official motivation was to revalorize traditional medicine, in decline due to British indifference, and also because it had traditionally been transmitted through a multiplicity of scattered lineages using esoteric languages[3] (WHO 2012, 71). Revalorization did not only mean preservation or restoration, but also improvement and diffusion, so as to render it as accessible, or more accessible, as its Western counterpart. This was particularly important, given that biomedicine was accessible to the entire population, and traditional medicine was purported to plug distributional gaps. Public health concern merges here with a certain nationalistic and anti-colonial ethos, much like it did in other countries in the region that also institutionalized their traditional medicine (Hsu 1999, 2001 on China; Wahlberg 2012 on Vietnam).

Another tacit motivation underpinned this project: to control and unify the country in service of nation-building. To understand this, we should situate the standardization and modernization of "Myanmar traditional medicine" within the historical and political context of a newly independent country where the rich ethnic and religious diversity is reflected in an equal diversity of medical traditions, and where the geopolitical structure is characterized by strong centralization and hierarchy. Since independence, the Burmese government has always tried to control and homogenize the peripheral regions inhabited by minority ethnic groups[4]: Rakhine, Chin, Kachin, Mon, Kaya, Karen, and Shan – some of which embrace religions other than Buddhism, which is the main religion of the Burmese and the official religion of Myanmar. The perpetration of control and homogenization has followed different routes, ranging from terror and internal displacements, to the spread of elements of Burmese Buddhism architecture over the national territory (Houtman 1999; Rozenberg 2001, 108, 116), and, I argue, the creation of a national medicine grounded in Buddhist culture. Indeed, despite the rich diversity of medical traditions belonging to the different ethnic and religious groups, for its "valorization project," the Burmese government only considered medical traditions from the Buddhist regions and neglected, or systematically excluded, those of the other minority ethnic groups. This means that the government did not intend to valorize a national medicine integrating the different traditions, but only the Buddhist strains among them. To maintain the illusion of a "national" medicine, and to mark its distinction from Western medicine, the medicine was named *taing-yin hsay* ("medicine of the country," or "indigenous/local medicine").[5]

Although built upon the contrast with Western biomedicine, as in other countries which encountered the same phenomenon (Adams 2002; Pordié 2011), the national medicine was shaped by biomedicine. Valorization occurs within the biomedical frame, at both the conceptual and the institutional levels. Despite representing an inimical and oppressive force, biomedicine was nevertheless valued as a symbol of modern science and a pathway toward the development and progress that the newly independent state aspired to. Perceived as the instrument through which traditional medicine would metamorphose from a scattered system embedded in irrational backward superstitions into something modern and scientific, the biomedical model was used as a reference to institutionalize and standardize traditional medicine.[6] Thus was established the Department of Traditional Medicine in 1989, a component of the Ministry of Health, which operated according to Western medicine criteria. Biomedical domination strengthened after the 1970s, when the project received the support of the WHO. Indeed, since the Alma-Ata Conference (1978), the WHO promoted the integration of traditional medicines in the national healthcare systems to plug biomedical shortfalls and facilitate its acceptance by local populations.

Perversely, the authority of the biomedical frame did not remove the nation-building motivations of the project. I contend that the ideals of science and modernity on which biomedical governance relies, as well as the standardization and disenchantment enforced by the different regulations, acquiesce with the state's nation-building agenda. The interplay and confluence between the two factors – the need to cope with the biomedical framework and nation-building aspirations – shaped the implementation of the project in its diverse dimensions: the regulation and standardization of teaching, and the practice and production of medicines.

That being said, it is important to remember that the valorization project occurred in a specific politico-economic context, where medicine in general – and traditional medicine in particular – represented the lowest priority on the state's agenda which was more focused on military affairs (Coderey 2016).

Institutionalization of the medical training and standardization of the curriculum

The first step to institutionalize traditional medicine involved the institutionalization and regulation of its teaching. In 1967, a teaching institute was opened in Mandalay, and another in Yangon. In 2001, a university in Mandalay replaced the two. Mandalay, the last Burmese capital before the arrival of the British, represented the core of Burmese Buddhist culture. By choosing Mandalay, the government highlighted the local, national character of the medicine, marking its difference from the "British" variety which had universities based in Yangon, the British capital. Moreover, given that Mandalay was also the center of Burmese Buddhist culture, this choice also expressed the domination of Buddhist culture over minority ethnic groups. However, despite the focus on local tradition, the head of the university, and a large part of staff in charge of its curriculum, were biomedical doctors.

A close look at the curriculum reveals an attempt to create a medicine which, while preserving traditional aspects, tries to comply with internationally accepted biomedical-based ideas of medicine; but which is also exempt from the elements perceived as potentially threatening for the state. The curriculum includes what is considered the "core of traditional medicine," Western medical science, and medical systems across Asia.

Representing "tradition" in the curriculum are four main topics: Buddhist medical knowledge (*betissa nara*), Ayurveda (*detissa nara*), astrology (*nakarta nara*), and alchemy (*weikzandara nara*). The first three systems provide the theoretical background for the diagnosis of natural disorders and their treatment by means of food and medical products that comprise ingredients of vegetal, mineral, and animal origin. The fourth, alchemy, differs in that its medicines are produced through the mixing and burning of metals, but is meant, once again, as a cure only for natural disorders.

Many traditional medical techniques used before and outside the formalized system have been omitted from the curriculum, and several of those selected have undergone a transformation; alchemy in particular. Alchemy is just one of four practices called *weikza* (literally "knowledge"), the others being the recitation of verbal formulas (*man*), and the distribution of small metallic sheets of paper inscribed with numbers or letters with Buddhist or cosmological meanings (*in* and *sama*) which the patient ingests or wears as amulets. Although *weikza* techniques were also used to cure natural diseases, they were primarily employed to protect and cure people from supernatural aggressions (*payawga*) emanating from spirits, witches, or sorcerers. Besides their therapeutic functions, they were also supposed to help the practitioner become a *weikza*: a super-powerful being able to exit the cycle of rebirth and access nirvana.[7]

That *weikza* practices have been excluded is unsurprising. They are incompatible with the Western conception of medicine, and out of tune with the image of modernity the state desires to exhibit to the world. They are, indeed, part of what the state sees as backward, irrational superstitions that keep the country in a premodern condition and, more importantly, corrupt Buddhism. Throughout history, state and religious authorities have conducted campaigns to eradicate those traditions and restore Buddhism to its claimed original purity (Houtman 1999; Sadan 2005; Rozenberg 2001) and rationality. Moreover, *weikza* practices are perennially perceived as a potential threat to state authority, primarily because they are disseminated in semi-secret congregations by charismatic personalities deemed to possess supernatural powers. Also, several congregations developed messianic and millenarist dimensions as they were centered on certain *weikza* believed to be the future Buddha or the king of the universe come to restore peace through respect of the Buddhist law (Ferguson and Mendelson 1981). On this account, state authorities have always tried to suppress these groups (Sadan 2005; Rozenberg 2001).

Against this backdrop, why has alchemy been retained in the curriculum at all? In response to a question about this, a highly ranking member of the university stated: "the TM council realized that it was important to keep alchemy because that too was part of TM and especially because they realized – and had proven – that it was really powerful" (traditional doctor, Mandalay, September 2014). If it is true that alchemy was part of medicinal heritage, the same holds true for the other *weizka* practices. I believe that as ingestible products obtained through the transformation of natural substances, alchemic products were closer to a Western idea of biopharmacology than mantras and diagrams. Moreover, unlike the other techniques that were used almost exclusively for the acquisition of extraordinary powers and protection against the supernatural, alchemy was also used to cure natural disorders, the only ailments the biomedical paradigm recognizes. I also suggest that by submitting alchemy to a standardized training, embracing Western science's view of heavy metals as toxic and dangerous, the state was hoping to reduce the risk of people being harmed.

The curriculum's second component includes subjects such as biology, physics, chemistry, botany, pharmacology, and anatomy. Their inclusion suggests a desire to draw traditional medicine closer to biomedicine and, as several doctors aver, to make it more systematic, precise, and scientific.

The third component includes traditional Asian medicinal techniques, with Chinese acupuncture and the Indian Ayurvedic massage occupying berths in the subject roster. The aim here is to enhance the legitimacy of Myanmar medicine through association with other well-established Asian medical traditions.

Despite the hybrid character of this medicine (Taylor 2005, 14), it is still presented as traditional; and despite the diverse origins of the different components, only the local roots and entanglement with Buddhism are underscored both in textbooks and discourses by the state's authorities (Kyaw Myint Tun 2001, 291; WHO 2012, 72).

Institutionalization and regulation of the practice

The second significant step toward state control over traditional medicine was its regulation and institutionalization.

To ensure that only medical knowledge disseminated in public institutes and circulated across the country would be practiced, the government enacted the Myanmar Traditional Medicine Council Law in 2000, under which only doctors who graduated from the public institutes, or external practitioners who attended a one-year special training in such institutes, were authorized to practice. In a further clampdown, in 2008, the law was amended to favor the university, which alone became the avenue to obtain medical licenses. Regulations were tightened further to prohibit traditional doctors from administering injections, performing operations, and attempting to cure cancers and the human immunodeficiency virus (HIV).

From the 1970s, several public hospitals and clinics have been opened countrywide, staffed by graduates from government institutes, most of who indulge in private practice to augment their income. The intent behind this vast swathe of medical cover is to provide healthcare to every citizen, and plug the lacunae in the biomedical structure. However, the rampant surge in the spread of a standardized Burmese-oriented medicine nationwide, including in minority-dominated areas where ethnic medical traditions exist, is as much a way to mark the dominance of the state over the subject as the installation of Burmese Buddhist pagodas in non-Burmese territories (Houtman 1999; Rozenberg 2001, 108, 116). It bears stating that although the state developed these services, it failed to provide them with decent financial contribution, leaving them in rather a poor condition.

The medicine practiced is largely a reflection of institutional teachings, with a focus on natural remedies and massages. Public services are provided with drugs manufactured by two government-owned factories in Mandalay and Yangon, whose units form annexes to the country's main traditional hospitals. There are 21 types of medicinal powders, produced according to a standard university manual, using material sourced from nine government-owned medicinal herb gardens located in Mandalay, Yangon, and Naypyidaw. Interestingly, astrology and alchemy are banned from the public healthcare facilities. Both are considered "non-scientific," and alchemy additionally potentially dangerous to human health because of the toxic component of its substances, as several doctors explain. It is intriguing that though the two techniques are taught at the university where they are perceived as medical heritage, their practice is proscribed. This suggests that the state's need to conform to biomedical criteria is stronger in the medical services largely because they are institutionally integrated into the global system of governance whose rules are monitored through a report system. Significantly also, the embrace of biomedical criteria is concordant with the image of the modern state[8] that the government wants to display. The public nature of these services increases their visibility and exposure, and hence the need for compliance.

Precisely because they lack such exposure, private sector services are minimally regulated and exempt from any explicit rule regarding the use of astrology and alchemy. Moreover, the large number of private healers, and the absence of a "reporting" system, decreases the possibility of a strict control system – especially in the semi-urban and rural areas.

Regulation of the production and circulation of medicines

The yen to improve the quality and safety of medical products by regulating the production and distribution of medicinal raw materials and powders on the one hand, and industrial ready-made products on the other, brought this sector under the regulatory and standardization ambit. Scrutiny of the implementation process reveals that it operates and circulates in accordance with biomedical legitimacy codes and criteria, but an attempt to reduce the regional cornucopia in the name of a standardized national medicine is also discernible.

If raw materials and powders have existed for eons – they actually represent an older form of traditional medicine – and have been accessible through traditional healers and vendors, the growth of national and, to a lesser extent, international trade has expanded their circulation and increased their accessibility. These products are sold in "traditional medicine shops" (*parahsay sain*) ubiquitous in urban and rural markets. Regardless of their location, all traditional medicine shops order their material from "retailer shops" (*hsay daik*) based in Yangon and Mandalay. These are supplied by two main sources: agents who collect materials from different parts of the country, and trading companies which import natural and chemical substances from different countries, mainly India, Bangladesh, Pakistan, Indonesia, China, and Singapore.

The government regulates and controls the quality and circulation of the products in two ways. First, it imposes a license to open such shops. A license is granted to retailers on payment, while proprietors of traditional medicine shops need to have some certified medical knowledge. As it is, most of these shops are run by families of traditional doctors. In its effort to standardize traditional medicine – ostensibly to control the quality, safety, and efficacy of the products – the Department of Traditional Medicine distributed the standard manual used in the public factories to every shop. The manual includes 57 medicinal formulas. The regulation and standardization of both the distribution and the combination of medicinal substances is intended to grant wider accessibility to good and safe medicines. It is undeniable also that blanket standardization supports the weakening of interregional, interfamilial, and interpersonal differences in the name of an emerging pan-national medicine.

Secondly, the government exercises control over the industrial and commercial sectors represented by public and private manufacturers. Medicines are produced as powders, balms, and tablets, and distributed in local and international markets. Raw material is obtained from retailers, private agents, and private medicinal herb gardens. Besides the aforementioned 21 medicines, the two government-owned factories also produce 12 kinds of tablets. Private corporations largely run the manufacturing industry, with most manufacturers hailing from traditional doctor families, or they are run by traditional doctors themselves. Most units were opened in the 1960s, with a numerical resurgence since 2010 in tandem with economic expansion. Large wholesale shops located in Yangon and Mandalay are the primary customers, from which all medicine shops, public and private, source their products. A small yet increasing number of companies also export their products to Malaysia, Singapore, Thailand, and China, where there is a Burmese diaspora.

In 1996, to "enable the public to consume genuine quality, safe and efficacious traditional drugs" (WHO 2012, 79), the government promulgated the Traditional Medicine Drug Law stating that drugs produced in the country have to be registered, and that manufacturers must possess licenses for their production. Licenses are issued only after products are checked and it is attested that they are produced in compliance with the standards of the GMP established by the WHO. Manufacturers send to the Department of Traditional Medicines samples of the raw material they use for their product, the final product, and the list of ingredients. In addition, products already in the market undergo occasional inspection. For export-oriented products, the manufacturer has to comply with the regulations of the importing country, and to register the product with the Food and Drug Administration (FDA). According to manufacturers and practitioners, the GMP guidelines are: the products must be natural; they should not be toxic or exceed the prescribed amount of chemicals or poisonous substances, such as heavy metals; they should include a limited number of ingredients; material must be of good quality; all ingredients need to be listed on the application form, as well as on the product's package. These regulations handicap manufacturers, given that traditional pharmacopeia often requires the use of many ingredients, sometimes even a whole plant. The norms completely exclude alchemical products from the market because they not only contain heavy metals, but are also prepared through secret formulas.[9] Although the industrial sector does not entail a standardization of the formulas, like it does for powders, it does impose a certain assimilation to

biomedicine through the simplification of the formulas and the exclusion of material unacceptable to biomedicine.

While procedures and controls are stricter for exports, as the concomitant risks
are much higher, manufacturers and doctors testify that controls are neither regular, nor efficient, in the domestic circuit. Officers in charge of trade and quality
control are eminently corrupt, and turn a blind eye to manufacturers with state or
army connections.

Another step to standardize the practice at the national and regional level is the
organization of national conferences. This comprises a conclave of all the members
of the association of traditional medicine practitioners, as well as the local representative of international meetings on traditional medicines, notably those organized by
the ASEAN. The participants, mainly Burmese from Yangon, are successful wealthy
manufacturers with special connections to the government. This affords further proof
that the traditional medicine regulatory project is very much a Burmese one.

All of the foregoing cumulatively suggests a governmental attempt to establish
and disseminate a new version of traditional medicine which meets the biomedical criteria of quality and safety, but which is free from regional specificity and
supernatural or esoteric components, perceived as a threat to its own power and
the modern image it aspires to. The ethnography of the implementation of the valorization project reveals how the need to comply with biomedicine supports these
political goals. The situation of Myanmar mirrors that described by Adams (2001)
and Janes (1995, 1999) in China, where applying scientific principles through
compliance with GMP allows and legitimizes the neutralization or elimination of
aspects in traditional medicine that the state perceives as anachronistic superstitions. A subtle difference exists between China and Myanmar in that religion is
the problematic component for the former, while anything non-Buddhist constitutes the concern in Myanmar. The situation is similar also to Vietnam (Thompson
personal communication; Wahlberg 2012), where the institutionalization process
helps the neutralization of ethnic differences and spurs the domination of ethnic minorities by the majority through the homogenization of the different forms
of medicines. This illustrates how local authorities employ global forms to meet
local political and economic agendas.

The implementation of the project is fraught with limitations. First, medicine is
low priority in the wider economic and political context of the state; and second,
the effete regulatory system cannot guarantee strict enforcement. Nebulous regulations, sectors and circuits less regulated than others, private sector vs. public
sector, and powders vs. industrial medicines open slats that manufacturers, sellers, and doctors jostle to their advantage, in much the same manner as Tibetan
medicine (Saxer 2013).

How do manufacturers, healers, and sellers – the targets of the state's imposed
regulatory regime – respond to and engage with the valorization project? The
ensuing section will examine in particular people's perceptions of the biomedical
dominance and nation-building agenda inherent in the project, and their reactions
to the neutralization of regional and lineage traditions and the consequent marginalization of the more spiritual and esoteric components of the practice.

My testimony will show that far from passively accepting or ignoring the regulations, manufacturers, healers, and sellers critically engage with them. Exploiting the semi-regulated nature of the system, they negotiate a space to maneuver. Their actions are guided by logic and values that differ from those of the state and which, more often than not, aim at preserving the diversity in medical efficacy and identity. How other forces, notably the market which operates through its own, purely commercial, logic comes to further contribute to this diversification trend forms part of the ambit of my testimony.

Complying or not complying – *or* an alternative form of governance

The majority of practitioners and manufacturers declare that the institutionalization and regulation of traditional medicine has been shaped by the biomedical frame, attributing this to the dominant position biomedical doctors occupy within the Health Ministry, and to the government's slant toward biomedicine rather than traditional medicine. They corroborate the subservience of traditional medicine to biomedicine with the fact that biomedical doctors head the university, biomedical subjects are rampant in the curriculum, and that alchemy, astrology, and injections administered by traditional medical practitioners are banned. While most agree that the standardization and regulation of the production of medicines is a systems-driven, scientific, and safe process, they also concur that it severely hampers the operation and organic expansion of traditional medicine. They infer from the prohibitions and hindrances imposed upon traditional medicine not only an expression of power from the state and biomedical authorities, but also a sign of mistrust in the abilities of traditional doctors.

There is less concern over the "nationalistic" thrust of the project, but rather more so with the apex position granted to biomedicine. It is acknowledged that the institutionalization project is fundamentally a Burmese operation, led by Burmese people, favoring Burmese practitioners, and disseminating the Burmese medical tradition at the cost of the others, but only a scant few discern the move as a part of nation-building. In the same vein, the attenuation of regional differences through homogenization of the practice, and the elimination of the more esoteric and spiritual components of it, are viewed against the biomedical yardstick rather than as an attempt of the government to neutralize aspects which could potentially threaten its authority and the image of a modern state.

The wide range of responses to the state's regulations needs acknowledgment. Actors exhibit different levels of compliance, depending on their position in the medical and political system, their medical or commercial aim, their ideological principles, and the material conditions in which they operate.

Practitioners of traditional medicine working in public services are the ultimate "products" of the institutionalization process, whose thoughts and actions are the most strongly shaped by the state regulatory system. Highly influenced by their training, they usually embrace the biomedical approach to herbal products. They generally perform their duties as ordained because they are subject to a reporting system, and because of the public exposure they attract. Their frustration stems from the limited resources the state allocates to them, which prevents them from providing quality service. A doctor from Yangon remarks: "The medicine we receive from the state factories are far from meeting the demand, so we often rely on products we buy from the market and we then combine them by ourselves." Ironically, where the practice is supposed to be the most standardized, much space is left to individual initiative. The same doctor complains:

> Our medicine has great potential; we could develop new medicines but for this we need to do research but we do not receive any money for it so we always work with the same medicines. . . . The problem is that the Ministry of Health is dominated by Western medicine; we are at the bottom of the hierarchy.

Many traditional practitioners display unconcern toward the prohibition on astrology and alchemy, and others register concern, while those who have private practices devise a way to employ these techniques.

Practitioners of traditional medicine who operate in the private sector are less restricted by regulations, and even when these apply, they are barely enforced. They use their freedom to practice according to their own principles. A great tendency to maintain individual, familial, and regional traditions exists, as well as a tendency to use practices banned in the public sector. This is mainly true for specialists trained outside institutions, who prescribe ad hoc medicines according to methods imbibed from their masters, using powders purchased at the market. This is, however, also true for doctors trained in state-run institutes, although they mainly rely on ready-made compounds in the forms of pills or powders. For instance, a registered practitioner from Thandwe tries to cure her patients by prescribing the standard powders, but resorts to adding other components if there is no improvement. She considers the characteristics of the patient, the weather, and often consults an ancient manuscript inherited from her grandfather. That she resorts to lineage tradition as the last recourse for difficult cases suggests that she considers this knowledge superior to the standardized, institutional model. The absence in the private sector of explicit rules that ban astrology and alchemy leaves the door ajar for the inclusion of these techniques, and many practitioners integrate them into their practice since they believe in their efficaciousness. Many practitioners, especially those who studied outside the university, use astrology as a complement for dietary and medical prescriptions, while many young university laureates consider astrology irrelevant. The use of alchemy is much rarer, especially among licensed healers. A practitioner explains: "it is a very difficult practice and at the university we just learn the basics; in order to learn more, one needs to have a further training with some 'external' masters, but these are becoming rare." Masters involved in this practice have usually learned the art from a master external to the state system, and operate in a clandestine manner. This is notably true for those who claim to cure HIV, which is specifically proscribed. Some, however, take more risks and sell their medicines from their homes, at the market, or even through other doctors.[10]

The standardization and centralization imposed by the government on traditional medicine has partially achieved what it set out to do. A clear sign of this is that people speak less in terms of regional medicine ("Rakhine medicine," "Shan medicine," and so on), and more in terms of national medicine (*"taing-yin hsay,"* "medicine of the country"). However, the standardization is countered by diversification and dispersing forces often tinged with resistance. As the standard manual is not obligatory, a certain freedom is left to the seller. If it is true that the majority of shops I have visited have plastic bottles labeled according to the mixture they contain, the contents of the bottles do not always correspond to what is prescribed in the manual. Many sellers do not strictly follow the manual, and try to improve the formulas by adding other materials based on their own medical knowledge and the characteristics of the patient, climate, and other factors. Many rely solely on family manuals passed down through generations, of which they are evidently very proud; remarks include: "ancient Rakhine," "from my grandfather who was a great traditional doctor," or "of my family who belongs to the lineage of doctors from the royal tradition."

If regulation of powdered medicines limits the variety of formulas, they do not affect the content of these formulas, at least not in a way that undermines the very principles of traditional medicine. This is actually what happens with industrial products. Most manufacturers are concerned since they believe that biomedicine-based regulations are "incommensurable" (Kuhn 1970) with traditional medicine, and that its application involves a significant transformation of the products and a loss of important aspects. An informant of Martin Saxer, over the application of GMP to Tibetan medicine, expressed it thus: it is like "sticking a goat's head on a sheep's body" (Saxer 2012, 498). In Myanmar too, many manufacturers believe that "it is the property of traditional medicine to use a plurality of ingredients, this is where the efficacy lies. . . and this is also how we neutralize toxic components" (interview with a manufacturer, September 2014).

Compliance with WHO guidelines and state regulations not only requires a certain acceptance of the submission to the biomedical ethos, but also involves economical resources. Money is needed to purchase and set up equipment that meets quality control requirements, apply for licenses, and pay for tests. Under the circumstances, very few manufacturers actually comply with the regulations. Those who do usually harbor export ambitions, since export/import controls are particularly strict. The most interesting case is Fame Pharmaceuticals. Founded in 1994, Fame produces mono-ingredient herbal medicines and food supplements by using organic material grown in special gardens. It has over 300 employees, including scientists, doctors, pharmacists, biochemists, botanists, microbiologists, and traditional medicine specialists. In January 2003, Fame received a GMP certificate from the Myanmar Ministry of Health, certification for its quality management systems from the International Organization for Standardization (ISO), and certification from Australian Certified Organic (ACO) of the Commonwealth of Australia. Thanks to the compliance with GMP guidelines, the circulation of Fame products has expanded across borders, but the efforts and requirements to meet that compliance have also rendered the products too expensive and largely inaccessible for the majority of the population. Moreover, Fame does not simply comply with GMP that is grounded in the biomedical framework, but follows methods in manufacturing its products through extraction from a single ingredient, which depart from the traditional polyherbal therapy. This closeness to biomedicine explains why Fame's goods are not sold in traditional medicine shops, but in pharmacies. As a result, even though the government presents Fame as *the* exemplary traditional medicine, the greater majority does not even see it as traditional medicine. Thus, compliance with biomedical criteria can lead to loss of the original identity and the acquisition of a new one.

Fame and its ilk are the exceptions that prove the rule. The greater swathe of manufacturers ignores or partially follows the regulations. Compliance is directed at achieving registration without tampering with the identity and efficacy of the medicines. I contend that the vast majority of the 2,578 registered manufacturers and 12,712 registered drugs (Ministry of Health Myanmar 2014) fall into this category. Many manufacturers use more ingredients than those allowed, including plants with toxic components or alchemical products containing mercury. Others enhance the efficacy of their products with pharmaceuticals. These additions are not declared to the control office, or mentioned on the label. A manufacturer interviewed in June 2014 explains: "If we list all the ingredients, everybody could reproduce our product; we want to protect our formula." If some of these additions are innovations, in the sense that these formulas do not exist in the traditional pharmacopeia, or fraud, when pharmaceuticals are added, others are attempts to produce an efficacious product. Some choose the middle path, complying with the regulations for market-borne products and bypassing them for products aimed for illicit circuits, in the manner described by Saxer (2013) with Tibetan medicine. While weak controls and corrupt agents facilitate the bypassing of regulations, the choice to do so is motivated by commercial and medical logic, tinged with defiance toward biomedical authorities. As stated by Quet et al. (2018, 15): "alternative routes, corrupt practices and other forms of arrangements are rather pervasive, sometimes inescapable and deeply embedded in official regulatory processes."

Despite regulations, many practitioners and manufacturers continue to resort to other traditional, mainly religious, practices to further increase the efficacy of their products, and hence ensure they retain their competitive edge. These practices have always been part of traditional medicine in the pre-institutionalized era, part of the larger repertoire that people use on a daily basis to increase the measure of fortune responsible for their well-being and the success of their endeavors.

The most common practice used to empower medicines is the recitation of Buddhist formulas (*gatha*) by doctors operating in both public and private sectors, primarily the former, and by manufacturers working in the industrial setting. Even the young business manager of Hman Cio I interviewed in June 2014 states: "one of the 11 *payeik* (protective formula said to be recited by the Buddha), the Bosenthout, is very powerful for health issues. While combining the powder I use to recite it or I switch on the CD of the recitation in the room next to the one I am working in!" Moreover, spiritual, supernatural, and mainly Buddhist power, alongside the presence of religious and devotional images, is channeled toward the medicines and to the medical practice. Indeed, today, as in the past, medicines are produced, distributed, and prescribed under Buddha's altar. In every house, clinic, hospital, and factory, the presence of the Buddha's altar grants protection and success to the medical practice.[11] The concept that the religious and spiritual status and attitude of the healer greatly affects his practice also remains very important. Even doctors who cure with ready-made compounds stress the fact that the efficacy of medicines greatly depends on the compassionate attitude of the practitioner and on his capacity to follow the Buddha's disciplines by respecting its main precepts, practicing meditation, and being involved in meritorious activities.[12] Hence, even though by acquiring compounded powders, traditional doctors participate in the standardization and disenchantment of the practice, they concurrently personalize these products by transforming them through their religious practices.

Drawing their efficacy from spiritual and other invisible forces, these practices remain largely untouched by the regulations which have focused largely on the tangible materiality of practices, the only one whose efficacy and safety can be scientifically proven. The threat in this case has come from the institutionalization itself, and the resultant transformations thereof. The mechanization and modernization of the production of medical products, and the separation between their production and distribution (the two actions are accomplished by different people in different spaces), have drawn medicines out of the ritualistic and religious context in which they were originally prepared and distributed. Despite this, these practices do not disappear, but adjust to the new production environment. The main change is that those practices emanate almost exclusively from Buddhism, and less from traditions that over time have come to be perceived as backward superstitions – and this is especially true of *weikza* and astrology. The transformation has to be attributed to the anti-superstition campaign the Burmese state and religious authorities have been conducting since the time of the royals to promote a purely Buddhist nation. Moreover, these changes are also in tune with the emphasis the state puts on the connection between traditional medicine and Buddhism.

The standardization and disenchantment of the medicine brought about by the institutionalization process is thus countered by healers and manufacturers who try to preserve its diversity and its spiritual and esoteric aspects. Other forces reach out to support this action, the market being the most relevant one. In fact, the market's very existence relies on diversity and visibility. It offers manufacturers the chance to capitalize on the traditional identity of their medicines by pushing them to create brands and logos, and to advertise through posters, billboards, TV and radio advertisements, and, very recently, websites. Names and logos of the brands, as well as images on the package and on advertising material, make use of traditional, cultural, and, in many cases, religious symbols. These appeal greatly to local and international clientele, who see them as sources of efficacy, authenticity, and legitimacy.[13] They are both a marketing strategy and a protection of traditional identity. Hence, by branding and advertising, the commercial sector contributes to the resistance to standardization promoted by the international guidelines imposed on the manufacturing industry and, more generally, the governmental project of formalizing and institutionalizing traditional medicine. Moreover, branding and advertising favor larger circulation. In other words, what truly emerges as national is no more a standardized version of medicine from Buddhist regions, as desired by the government, but a plurality of medical traditions belonging to different regions and lineages.

Conclusion

A scrutiny of the governance of traditional medicines in Myanmar reveals two distinct forms, the one implemented by the state, oriented toward standardization; the other operated by manufacturers and healers, supported by the market, and oriented toward diversification, as attested by Quet et al. (2018, 5) elsewhere in Southeast Asia. The two are organically interlaced in that they operate in relation and reaction to each other. Although both are informed by logics of similar nature (medical, political, economic, and identity), their equations with those logics are often at variance.

The examination yields an understanding of the political motivations behind the institutionalization project, and the social forces which concurred to shape it. I have shown that the regulations implemented are largely the outcome of a local re-appropriation of biomedical norms with the two-fold agenda of compliance with global medical expectations, and nation-building. Indeed, the standardization and modernization of the practice contributed to the neutralization of individual, ethnic, familiar diversity, and the dilution of the practice, in the name of building a modern united Buddhist nation. However, its actual implementation is fraught with limitations in the context in which it unfolds, wherein traditional medicine occupies a lower position within the government agenda. Moreover, regulations are not applied evenly, as the implementation largely depends on the geopolitical and hierarchical structure of the country, the level of exposure each sector has to external global authorities, the degree of efficiency of all bureaucratic and security services – specifically officers who are in charge of checking that manufacturers and healers comply with regulations, and either enforce them or otherwise. The gray area created by ambiguous regulations, and the inherent weakness of the regulatory system, leaves room enough for traditional diversity to subsist, although in a renovated form.

Manufacturers and healers perceive not so much the nation-building project, but the dominating power of biomedicine endorsed by the state. Hence, their response is not just to accept or oppose the different ideas on producing medicines, but of taking a political stance. The manner of response varies according to the specific position actors occupy in the medical and social space, their material constraints, and their market ambitions. The responses range from partial adaptation and compromise with the aim to foster circulation (Adams 2001, 2002; Adams et al. 2010; Wahlberg 2012; Craig 2012; Saxer 2013), to circumvention of the rules by taking advantage of the haze of the regulatory regime (Saxer 2013) with the aim of preserving traditional diversity, specificity, and efficacy. The majority seems to take a middle stance, endeavoring to strike a balance between expanding its circuits and preserving its identity, supported in this by the diversifying force of the market.

Notes

1 The main locations for this fieldwork were Mandalay, in Central Myanmar; Yangon, in Lower Myanmar; and Thandwe, in the central part of Rakhine State. Mandalay and Yangon are the main cities of the country and host the main institutions of traditional medicine, while Thandwe is a rural area of a peripheral state where the development of this medicine is still very limited. Several fieldwork trips have been conducted in these three areas between 2011 and 2016, with significant financial support from the Asia Research Institute of the National University of Singapore. Information was gathered through observations and semi-structured interviews with specialists of traditional medicine (doctors, vendors of medical products, owners and managers of medicines companies) and lay people, consumers and patients. I am deeply grateful to Jeremy Fernando for helping me thinking through this chapter and for editing the language.

2 In 1989, the military government changed the name of the country from Burma to Myanmar. Yet "Burmese" is still the name more commonly used to refer to the population of the country, regardless of the ethnic belonging.

3 The neglect of traditional medicine on the part of the British is a rhetoric created and perpetuated by the Burmese government. Actually, since 1928, the British government started envisioning the possibility of a certain integration of "indigenous medicine" into the health system and started planning for it (Aung-Twin 2010). Although this project was never realized, it did provide a basis for the Burmese project. Indeed, the way in which the Burmese government later formalized this medicine was largely based on the British plan. A similar rhetoric of neglect – motivated by similar reasons – is attested also in Vietnam by Wahlberg (2012).

4 Soon after independence, the military junta assumed power and retained it until 2011, when a semi-civil government was established. The democratic transition was completed in 2015 with the victory of the democratic party in the national election.

5 In the official translations in English, the expression which is used is "traditional medicine," this being the term used by global health actors (WHO *in primis*), which, at the same time, has the advantage of highlighting the local and ancient roots of this kind of medicine.

6 That being said, it is also true that as suggested by Wahlberg's (2012) works on Vietnam, and Hsu's (1999, 2001) and Saxer's (2013) on China, the institutionalization and modernization of traditional medicine are not necessarily the expression of biomedical and Western colonization, but are very much part of the local bureaucratization system in line with the modern development agenda of the state.

7 The term *weikza* refers at the same time to the practices, the knowledge, or, rather, the power these practices grant access to, and the status one reaches when acquiring that knowledge.

8 It was in the same spirit of displaying conformity and exhibiting an image of modernity, that the government eliminated statues of *weikza* and guardian spirits from tourist areas (Sadan 2005, 109).

9 This contrasts with the case of the *tsothal* described by Saxer (2013, 71–75). *Tsothal* is the main ingredient of the famous Tibetan "precious pills"; it contains purified mercury and gold, and its preparation includes complex rituals combining meditation and chanting. Despite its strong religious component, which is problematic for the Chinese state, and despite its containing mercury and gold, in opposition to Chinese GMP and drug laws, Chinese regulations do not impinge on its preparation. *Tsothal* is recognized as

part of Tibetan cultural and scientific heritage and, since 2006, as part of China's intangible cultural heritage. That said, following the introduction of GMP in China in 2001, it was declared that by 2004, *tsothal* must be prepared according to GMP guidelines. Yet, Saxer affirms that nothing much has changed so far.

10 Particularly striking is the case of Hsaya Gyi U Shein, the most well-known alchemist of the country who passed away in 2014 at the age of 89. For more than 30 years, he had been dispensing treatments in his house in Yangon, using alchemic medicines. His wide renown at both the national and international levels, and the possession of diplomas from foreign institutes, granted him legitimacy – and protection – that he was lacking from the state.

11 This brings to mind Saxer (2013, 189), who reports of specialists of Tibetan traditional medicine decorating their studios with *tangkas* and religious images.

12 The daughter of the owner of Hman Cio I have interviewed in June 2014 stresses that every year her brother sponsors the initiation ceremony of several children and that the family, like other families of successful manufacturers, regularly organizes donation ceremonies to monasteries or donate free medicines to monks, nuns, and orphans. She also adds that annually she herself "leaves the world" for a period of a month to become a nun and meditate. If in their eyes, these practices increase the family karma and hence the efficacy of the medicines it produces, it is undeniable that they also contribute to the company's social respectability and popularity. Even if this outcome is not stated it is certainly part of the agenda, given that these events are always publicized, and pictures displayed on the walls of the shop.

13 In most cases, these aspects remain invisible to the public, all the more so given the fragmentation of spaces and separation of roles; to customers having access only to the final product, the production process remains veiled. If manufacturers do their utmost to publicize these aspects, the market offers the opportunity to render them even more visible.

References

Adams, Vincanne. 2001. "The Sacred in the Scientific: Ambiguous Practices of Science in Tibetan Medicine." *Cultural Anthropology* 16 (4): 542–575.

Adams, Vincanne. 2002. "Establishing Proof: Translating 'Science' and the State in Tibetan Medicine." In *New Horizons in Medical Anthropology: Essays in Honour of Charles Leslie*, edited by Margaret Lock and Mark Nichter, 200–220. London: Bergin and Garvey.

Adams, Vincanne, Mona Schrempf, and Sienna Craig, eds. 2011. *Medicine, between Science and Religion*. New York: Berghahn Publishers.

Aung-Twin, Maitrii. 2010. "Healing, Rebellion, and the Law: Ethnologies of Medicine in Colonial Burma, 1928–1932." *Journal of Burma Studies* 14 (1): 151–186.

Burawoy, Michael, Joseph A. Blum, Sheba George, Zsuzsa Gille, and Millie Thayer. 2000. *Global Ethnography: Forces, Connections, and Imaginations in a Postmodern World*. Berkeley: University of California Press.

Coderey, Céline. 2016. "Accessibility to Biomedicine in Contemporary Rakhine State." In *Metamorphosis: Studies in Social and Political Change in Myanmar*, edited by Renaud Egreteau and François Robinne, 260–287. Singapore: NUS Press.

Collier, Steven J., and Aihwa Ong. 2005. "Global Assemblages, Anthropological Problems." In *Global Assemblages: Technology, Politics, and Ethics as Anthropological Problems*, edited by Steven J. Collier and Aihwa Ong, 3–21. Malden, MA: Blackwell.

Craig, Sienna. 2012. *Healing Elements: Efficacy and the Social Ecologies of Tibetan Medicine*. Berkeley: University of California Press.

Ferguson, John P., and E. Michael Mendelson. 1981. "Masters of the Buddhist Occult: The Burmese Weikzas." *Contributions to Asian Studies* 16: 62–80.

Foucault, Michel. 1991. "Governmentality." In *The Foucault Effect: Studies in Governmentality (with two Lectures by and an Interview with Michel Foucault)*, edited by Graham Burchell, Colin Gordon, and Peter Miller, 87–104. Chicago: University of Chicago Press.

Foucault, Michel. 1995 [1975]. *Discipline and Punish: The Birth of the Prison*. New York: Vintage Books.

Houtman, Gustaaf. 1999. *Mental Culture in Burmese Crisis Politics: Aung San Suu Kyi and the National League for Democracy*. ILCAA Study of Languages and Cultures of Asia and Africa Monograph Series, n° 33. Tokyo: Institute for the Study of Languages and Cultures of Asia and Africa, Tokyo University of Foreign Studies.

Hsu, Elizabeth. 1999. *The Transmission of Chinese Medicine*. Cambridge: Cambridge University Press.

Hsu, Elizabeth. 2001. *Innovation in Chinese Medicine*. Cambridge: Cambridge University Press.

Janes, Craig R. 1995. "The Transformations of Tibetan Medicine." *Medical Anthropology Quarterly, New Series* 9 (1): 6–39.

Janes, Craig R. 1999. "The Health Transition, Global Modernity and the Crisis of Traditional Medicine: The Tibetan case." *Social Science & Medicine* 48 (12): 1803–1820.

Kuhn, Thomas. 1970. *The Structure of Scientific Revolutions*. Chicago: University of Chicago Press.

Kyaw Myint Tun. 2001. www.searo.who.int/entity/medicines/documents/traditional_medicines_in_asia.pdf.

Langford, Jean. 2002. *Fluent Bodies: Ayurvedic Remedies for Post-Colonial Imbalance*. Durham and London: Duke University Press.

Ministry of Health Myanmar. 2014. www.asienhaus.de/archiv/asienhaus/kattermann-stiftung/gesundheitswesen/MOH-Health_in_Myanmar_2014.pdf.

Pordié, Laurent. 2010. "The Politics of Therapeutic Evaluation in Asian Medicine." *Economic and Political Weekly* 45 (18): 57–64.

Pordié, Laurent. 2011. "Savoirs thérapeutiques asiatiques et globalisation." *Revue d'Anthropologie des Connaissances* 5 (1): 1–12.

Quet, Mathieu, Laurent Pordié, Audrey Bochaton, Supang Chantavanich, Niyada Kiatying-Angsulee, Marie Lamy, and Premjai Vungsiriphisal. 2018. "Regulation Multiple: Pharmaceutical Trajectories and Modes of Control in the ASEAN." *Science, Technology & Society* 23 (3): 1–19.

Rozenberg, Guillaume. 2001. *Thamanya: enquête sur la sainteté dans la Birmanie contemporaine*, PhD Diss. École des Hautes Études en Sciences Sociales, Paris.

Sadan, Mandy. 2005. "Respected Grandfather, Bless This Nissan: Benevolent and Politically Neutral Bi Bo Gyi." In *Burma at the Turn of the 21st Century*, edited by Monique Skidmore, 90–111. Honolulu: University of Hawaii Press.

Saxer, Martin. 2012. "A Goat's Head on a Sheep's Body? Manufacturing Good Practices for Tibetan Medicine." *Medical Anthropology* 31 (6): 497–513.

Saxer, Martin. 2013. *Manufacturing Tibetan Medicine: The Creation of an Industry and the Moral Economy of Tibetanness*. New York and Oxford: Berghahn.

Taylor, Kim. 2005. *Chinese Medicine in Early Communist China, 1945–63: A Medicine of Revolution*. London: Routledge.

Van der Geest, Sjaak, Susan Reynolds Whyte, and Anita Hardon. 1996. "The Anthropology of Pharmaceuticals: A Biographical Approach." *Annual Review of Anthropology* 25: 153–178.

Wahlberg, Ayo. 2012. "Family Secrets and the Industrialisation of Herbal Medicine in Postcolonial Vietnam." In *Southern Medicine for Southern People: Vietnamese Medicine in the Making*, edited by Laurence Monnais, Michele Thompson, and Ayo Wahlberg, 153–178. Newcastle upon Tyne: Cambridge Scholars Publishing.

WHO. 2012. "Traditional Medicine." www.searo.who.int/entity/medicines/topics/traditional_medicine_in_myanmar_2012.pdf.

Whyte, Susan Reynolds, Sjaak Van der Geest, and Anita Hardon. 2002. *Social Lives of Medicines*. Cambridge: Cambridge University Press.

4 Negotiating Chinese medical value and authority in the (Bio)polis

Arielle A. Smith

At the opening of the Strait of Malacca – the most direct shipping channel connecting the Pacific and Indian oceans – Singapore has long been a node through which people, goods, and ideas circulate. The main island's complex trading history has linked various parts of the region and beyond for centuries, from its entrepôt role in the historic maritime routes of the Silk Roads and their proposed reinstitution in China's Belt and Road Initiative, to 20th-century investment in regional Special Economic Zones (e.g., China's Shenzhen Industrial Park and the Singapore-Malaysia-Indonesia Sijori growth triangle). Leveraging a carefully cultivated "multiracial" population (touted as Singapore's only natural resource), the post-colonial state has positioned Singapore as "the hub of an 'effervescent ecosystem' of global capital," within an emerging, synergistic network of regional and global flows (Ong 2006, 26).

Built upon the "modern," high-tech nation-state proposed by the state's Intelligent Island strategy (Clancey 2012), Singapore's biopharmaceutical industry has been marketed to attract international investors, multinational corporations, and biomedical professionals. As a hub for research and development, Singapore's Biopolis complex is therefore not only the name of a technology park in which public and private interests converge to produce marketable products; it also materializes a singular, "technotopian" vision for the bioeconomic development of Singapore and its perfectible body politic (Waldby 2009). Plato's well-ordered city-state, or polis (as debated in the *Republic*), is thus reimagined in an increasingly transnational geopolitical landscape and filtered through the lens of biopower. Grounded in biomedical epistemology, standards, and values, these modernist processes have decidedly excluded "traditional" practices like Chinese medicine,[1] which nonetheless continues to assemble/ reassemble in relation to Singapore's ethnic Chinese majority and emerging transnational markets.

My primary exploration of Chinese medicine in Singapore – between January 2006 and October 2007 – led me all over the main island and, on one occasion, to the nearby Indonesian island of Batam. Most of my fieldnotes pertain to daily life, clinical procedures, and embodied experiences of an eclectic group of Singaporeans, the majority of whom were of ethnic Chinese descent. In addition to daily observations of high-rise apartment life, public transportation, hawker centers, and other public spaces, I was fortunate to conduct long-term participant observation with several full-time, highly esteemed Chinese medical physicians in charity clinics and private practices. I also interviewed shopkeepers, factory managers, researchers, and Ministry of Health officials, attended public health events and lectures, toured Chinese medical factories and research facilities, and conducted observations and interviews at a popular Chinese medicinal herb and food shop.

Toward the end of my first year of fieldwork, I was introduced to Dr Song, a Chinese medical physician who worked with the Chinese food and medicine chain Hock Hua (also referred to as Fu Hua, in Mandarin). Like many other Chinese-stream educated Singaporeans, Dr Song explained, he became interested in Chinese science (i.e., scientific research and innovation conducted in China) and medicine after the opening of the People's Republic of China (PRC) in the late 1970s. In particular, he was inspired by PRC-produced documentaries that highlighted Chinese aerospace engineering and acupuncture analgesia. After his training at the Singapore College of Traditional Chinese Medicine (SCTCM), Dr Song took a medical sales position with the British trading company Jardine Matheson for eight years, and then pursued a business degree at the University of Melbourne. Throughout this time, he continued to practice Chinese medicine in private clinics and, after returning to Singapore, volunteered at Chunghwa Yiyuan (the charity clinic affiliated with SCTCM). After working as a Chinese medical consultant for several other companies, Dr Song joined Hock Hua in 2000.

With over 20 years of experience, Dr Song advised Hock Hua on matters pertaining to the import, processing, packaging, sale, and export of Chinese medicinal products. He also maintained a private clinical practice and occasionally gave public lectures or media interviews on topics related to Chinese food and medicine. In his capacity as a physician, he navigated the relatively new regulation of clinical practice (i.e., the Traditional Chinese Medicine Practitioner's Act of 2000, revised and implemented in 2001); as a consultant for Hock Hua, he also followed developments in manufacturing and import practice. Both tasks required a nuanced understanding of the changing social and regulatory position of Chinese medicine in not only Singapore, but also in the PRC, Southeast Asia, and to some extent in Europe and North America.

In so far as biomedical epistemology, theory, and practice are dominant in Singapore, physicians, entrepreneurs, and researchers alike have made increasing efforts to reframe Chinese medicine with reference to biomedical standards and values. Dr Song traced this effort at reframing back three generations: while first generation (colonial era) Singaporean Chinese did not generally question Chinese medicine, by the third generation, many began to reject Chinese values and practices. Under the split-stream education system, he explained, second generation (mid-20th-century) Chinese-stream Singaporeans often suffered economically and socially by comparison with their English-stream counterparts.[2] Hence, third generation ("post-colonial") Singaporean Chinese were encouraged by their parents to study English and, as he put it, "absorb Western values," effectively marginalizing "traditional" practices like Chinese medicine. "They use Western eyes – Western-trained eyes – to look at things," Dr Song observed. "That sometimes makes things very difficult for us." Nonetheless, he insisted, it was the Chinese medical community's responsibility to demonstrate their suitability as a mainstream medical practice in "modern" Singapore.

One of the strategies used by Hock Hua to negotiate their status in Singapore involved "modernizing" their stores with reference to contemporary interior design standards: lighter colors, glass shelving, tidy new displays, and more lighting fixtures. These aesthetic changes not only harmonized Hock Hua's retail outlets with Singapore's urban landscape, but also mirrored earlier efforts in the PRC, such as those undertaken by Great Five Continents Drugstores in the early 20th century. In contrast with "traditional" Chinese shops – typically one or two dimly lit stories with wooden fronts, no windows, overhanging eaves, lacquered wooden shop signs, and cloth shop symbols – Five Continents shops were two or more brightly lit brick or concrete stories with "Western-style" decorations, eye-catching interiors, windows, and wide entrances (Cochran 2006). Although these shops largely sold "Western-style" goods, while Hock Hua primarily sold Chinese foods and herbs, both accommodated local design tastes and Euro-centric notions of modernity in their appearances.

Competing for scarce resources and medical authority in Singapore, some companies and individuals engaged in discourses of efficacy, standardization, and "modernity," seeking to elevate Chinese medical practice and products. Others appealed to history, collective medical experience, and/or Chinese "tradition," evoking cultural values initially rejected and then briefly promoted by the postcolonial state. Still others promoted Chinese medicine as "complementary and alternative medicine" (CAM) – a framework imported from the United States and Europe and adopted by the Singaporean government. According to Dr Song, this last strategy particularly appealed to younger Singaporeans, who sought "natural" or "alternative" remedies, generating a niche market in which even the government saw potential. More recently, Singapore was positioned as a regional clearinghouse for herbs and other products cultivated and/ or processed in the PRC, because their reputation for stringent safety and quality controls (e.g., ensuring authenticity and consistent grading) alleviated fears of contaminents and counterfeit products that circulated with the flow of goods out of the PRC.

Although far from exhaustive, these strategies allude to a variety of dynamics that Chinese medical professionals negotiate as they vie for authority and security in increasingly global medical markets. Like the PRC, Singapore evaluated Chinese medicine against biomedicine in the course of 20th century development. Unlike the PRC, biomedicine was selected as the sole authority for medical services and standards in Singapore. At the confluence of charitable and commercial work, Chinese medicine thus emerges in somewhat strained relation to Singapore's biopolitical processes. Appraised against an exclusively biomedical healthcare system, this mercurial assemblage is often framed as a "complementary" practice with economic potential; at worst, it is sometimes depicted as antiquated and "unscientific" quackery.

In order to properly situate this status disparity, I will first briefly discuss the historical development of the nation-state, nationalism, and governmentality, as they pertain to post-colonial nation-building in Singapore. I will then review the politically motivated formulation of "traditional Chinese medicine" (TCM) in 20th century China, and its conscription to the service of Singaporean biopower at the turn of the century. Against this sociopolitical backdrop, I will describe various convergences and divergences of biomedical and Chinese medical practices highlighting, first, "healthy lifestyle" and "nurturing life" practices and, second, techno-scientific discourses of safety and authenticity surrounding Chinese *materia medica*. Finally, I will consider the interplay of governance and circulation with respect to biomedical standards, the harmonization of "traditional medine" regulatory frameworks, and Singapore's nationalist narrative. In this manner, I will explore several ways in which the 20th- and 21st-century transformation and circulation of Chinese medicine has been modulated in accordance with dynamic sets of interests. This chapter will therefore illustrate how fluidity, permeability, and complex power relations characterize both the governance of Singapore's post-colonial sociopolitical milieu and the practice of Chinese medicine therein.

Colonial and post-colonial biopower

Contemporary Southeast Asia, as a geopolitical entity – represented on maps and charts as a collection of discrete nation-states – was largely sculpted by colonial and post-colonial processes. At the end of the colonial era, power relations between colonizer and colonized (North and South) were renegotiated through discourses of modernization and development, and then neoliberal globalization. While aspects of Euro-American political economy and "modernity" were contested in this tumultuous period (and thereafter), the notion of the nation-state and its associated art of government endured in many places as the region was carved into discrete territories.

To some extent, the birth of the nation-state and nationalism was preceded by a shift from sovereignty to government rationality, or "governmentality" (*raison d'État*), as described by Michel Foucault (1991). Governance (or sovereignty) during the time of Machiavelli was largely a matter of identifying potential threats to the sovereign's territory rather than a concern with the people who occupied it. By contrast, the goal of governmentality became the surveillance and control of individuals, goods, and wealth (economy). Europe's demographic transition of the 18th century produced a "problem of population" and a new science of governance was created, referred to as political economy. This approach shifted the plane on which economy was conceived from the family to the population, management of which required data collection, analysis, and future projections (i.e., statistics). The concept, body of knowledge, and practices of political economy therefore developed within the context of European governmentality, as the central concerns of government shifted from the maintenance of territory to the management of men and things.

By the end of the 18th century, the sovereign's right of death over his subjects was replaced by governmentality's emphasis on the management of the life of populations; this power to manage life took two complementary forms. Foucault

1990 ([1976]) refers to the first technique of power as anatomo-politics, or procedures and practices that treat institutions (and bodies therein) as machines, disciplined to optimize their productive capacities. In short, the discipline of individual bodies within educational, military, or medical institutions ensured a docile and productive population that fueled economic development. The subsequent approach concerned governmental intervention and regulatory control of biological processes such as propagation, health, and mortality, which he refers to as biopolitics. While these two techniques of power developed independently, once conjoined at the end of the 18th century they marked the beginning of an "era of biopower": "the great technology of power in the 19th century," indispensable to the development of capitalism (140).

In addition to a disciplined and regulated labor force – vital for industrialization and the development of capitalism – a loyal body politic (i.e., sense of nationalism) was also essential to developing nation-states. As Hutchinson and Smith (1994), Eric Hobsbawm (1983), Benedict Anderson (1991), and others have argued, the formation of many nation-states relied on the presence, creation, or consolidation of ethnic communities around which the state could develop loyalties. In the event this ethnic solidarity did not already exist, or was fragmented between multiple communities with competing territorial claims, it had to be created. Nationalism, in this context, can be understood as one of the ideological goals of the state: a sense of loyalty or affinity to a particular nation that offers popular freedom, sovereignty, and fraternity (or solidarity), and thus a common identity (Hutchinson and Smith 1994, 4–5). Discourses of the nation-state and governmentality spread quickly throughout Europe, European colonies, and beyond. In the postcolonial era, newly sovereign nation-states in Southeast Asia thus negotiated Euro-American notions of "modernity" and governmentality, on the one hand, and the "traditional" knowledge, practices, and values of their indigenous and migrant communities, on the other.

Prior to colonization, most Chinese in Southeast Asia traveled for trade-related purposes; relatively few stayed in Singapore for an extended period of time. Chinese merchants and seafarers, such as Zheng He in the 15th century, were prominent agents in the regional circulation of goods via maritime routes of the Silk Roads. Select Chinese *materia medica* and foods like birds' nests and agarwood – sourced from South and Southeast Asia – were imported to China, while silk, lacquer ware, porcelain, and other Chinese goods were carried to Europe via India. After Singapore was established as a free port by the British East India Company in 1819, its tremendous economic success ushered in several waves of Indian and Chinese migrant labor. The pre-colonial flow of commodities in the region thus merged with an unprecedented circulation of people in the colonial era. Migrating first from Malacca and Penang in contemporary Malaysia (at that point referred to as "Malaya"), and then directly from southern China, Chinese laborers and merchants constituted a numerical majority by 1849.[3]

Under the British East India Company, and as a crown colony, practices and loyalties associated with Chinese heritage were permitted in so far as they did not interfere with the administration of the colony. Healthcare for Singapore's overseas Chinese communities was largely provided by Chinese medical physicians who set up services through native place associations, often offering their expertise free of charge (Quah 1989, 132–133). Chinese medicine was not seen as a threat – or even an issue that warranted attention – by the colonial authorities. By the mid-20th century, however, political developments in China and Southeast Asia (including regional anti-colonial sentiment and concerns about Chinese nationalism) discouraged the overt promotion of Chinese heritage in Singapore (Kong 2003, 64). Furthermore, the Malayan identity promulgated at the end of the colonial era was soon curtailed by separation from the Federation of Malaysia in 1965. In place of ethnicity-based solidarity, a "multiracial" Singaporean nationalism was developed, and a "docile," productive population was engineered in order to attract foreign investment and fuel economic development (Grice and Drakakas-Smith 1985, 348).

A history of colonial subjugation, a failed Malayan identity, and the unrest promised by promoting Chinese culture in its place were all seen as antithetical to the ideal Singaporean national identity. Thus, history and heritage were carefully managed by the state, and actively promoting Chinese cultural practices such as Chinese medicine became politically untenable for over three decades. Since the late 1970s, however, Singapore's political relationships with its neighbors and the PRC have stabilized, permitting recognition of the economic and moral value of Chinese language and culture by the turn of the century. Playing on Singapore's strengths as an entrepôt economy, the increased circulation of goods, ideas, capital, and people expanded the possibility of cultural growth and expression in Singapore, and perhaps motivated the state to draw practices such as Chinese medicine under more direct surveillance and control.

Traditional Chinese medicine: enclosed, partitioned, and disciplined

The model upon which the post-colonial Singaporean state explicitly based its governmentality was inherited from the British, and, although they later chose to promote "Asian values" such as Confucianism, they retained this fundamental approach throughout. Thus, similar techniques of power as those previously described were employed to transform small, heterogeneous, and economically unstable Singapore into a "First world" capitalist nation within three decades. Biomedical research and development was encouraged as an important economic growth sector – designed to attract international investors, biopharmaceutical companies, and biomedical professionals, and to promote Singapore's image as a "modern," high-tech nation-state. A productive body politic – surveilled, disciplined, and categorized – was engineered as the country's primary natural resource and embodiment of the "Biopolis of Asia."

While the 21st-century emphasis on biopharmaceutical research and development – one pillar of an increasingly diversified economy – had little room for Chinese medicine, the transition from "Intelligent Island" to "Biopolis" (Clancey 2012) was roughly coterminous with increased regulation of Chinese medicine. Guided to some extent by the ongoing regulation, standardization, and "modernization" of Chinese medicine in the PRC, Singapore's strategy differed in this unambiguous deference to biomedicine. In order to explore these congruities and incongruities, this section will briefly review the 20th century reformulation of Chinese medicine as TCM in the PRC and its subsequent regulation in Singapore at the turn of the century. I will then illustrate how daily clinical practice in Singapore can be understood with reference to biopolitics and anatomo-politics (biopower).

The theory, practice, and epistemology of Chinese medicine developed heterogeneously, as reflected in two millennia of written commentaries, treatises, and reinterpretations. Ralph Croizier (1976) explains that factions within the post-Imperial Republic and the May Fourth Movement sought to eradicate Chinese medicine as an ancient symbol of Chinese culture, while others sought to conserve it. Subsequently, prominent voices in China sought to preserve Chinese medical identity as an expression of both progress and patriotism, by organizing and regulating practitioners. This ideological promotion of the so-called "medical legacy of the Motherland" culminated in the compilation of folk remedies as proof of the people's medical wisdom during the Great Leap Forward, but dwindled thereafter due to economic and political pressures. Finally, during the Great Proletarian Cultural Revolution, advocates once again took up the cause of preventing the dissolution of Chinese medicine by reorganizing medical schools to integrate Chinese and "Western" medicine.

Meanwhile, Paul Unschuld (1985) frames the state's management of Chinese medicine within a longer history of social metamorphosis, asserting that the post-Imperial management of Chinese medicine was the third major attempt at its legitimization. This was preceded, he explains, by the much earlier shifts from magic-based remedies to pragmatic herbal remedies, and then to pharmacological doctrine. Once its foundation in systematic correspondence (Porkert 1974) and imperial worldview was removed in the mid-20th century, a new form emerged that reflected Marxist-Maoist values. Elisabeth Hsu (1999) also notes the discontinuity between contemporary Chinese medical textbooks and long-standing concepts and practices of Chinese medicine. Her ethnographic account of the transmission of Chinese medicine illustrates how 20th century "TCM" textbooks were reframed according to political ideology and "Western" science, in order to produce biomedically-oriented theory in place of Chinese medical doctrine.

Finally, Volker Scheid (2002) insists that Chinese medicine is not a totality and thus cannot be reduced to a singular cultural logic or process. Rather, he claims, it is currently emerging and disappearing simultaneously, subject to global synthesis and local production. Using the term "Chinese medicine" to refer to the scholarly elite practice of the Imperial, Republican, Maoist and post-Maoist periods, he notes that physicians frequently incorporate biomedical diagnoses and prescriptions into their practice, while biomedical doctors in China may use Chinese *materia medica*. Scheid describes how physicians' practices must accommodate patients' experience (and terminology) in their symptomatic descriptions, and suggests that this "grassroots pressure" has the power to collectively shape the practice of Chinese medicine (122–128).

Chinese medical theory and practice have been significantly transformed over the course of several thousand years, particularly in response to dramatic changes in political economy during the 20th century. By the turn of the century, this standardized form of Chinese medicine ("TCM") became the only "traditional" medicine accommodated within Singaporean biopolitics at all. Building on the professionalization efforts already undertaken by the Chinese medical community in Singapore, and with reference to those in the PRC, several pieces of legislation were enacted at the turn of the century that drew Chinese medicine further into the domain of biopower. This regulation, overseen by the Ministry of Health (MOH), politically legitimated only a portion of Singapore's varied Chinese medical community.

Since the early colonial era, Chinese medicine has been practiced in Singapore by herbalists, acupuncturists, food and medicine shopkeepers, bonesetters, physicians, and others. Their treatments include herbs at various stages of processing (bulk, powdered extracts, liquids, pills, and so on) to be taken orally or used as compresses, topical oils, acupuncture, moxibustion, massage, cupping, bloodletting, scraping, and qigong – to name the most common. Although some Chinese medical physicians have been known to make house calls, either independently or as part of a service, most operate in Chinese medical halls, institutes, food and medicine shops, dispensaries, and "charity clinics." With no hinterland for agricultural production, most Chinese *materia medica* and instruments (such as acupuncture needles) are imported from Taiwan, Hong Kong, or mainland China, with the notable exception of materials sourced in Southeast Asia.

For example, swallows' nests (*yanwo*) have been luxury trade items that have circulated between southern China and Southeast Asia since the 16th century, although they are now predominantly exported to Singapore, Indonesia, Thailand, Taiwan, and Hong Kong (Leh 1993). Traditionally harvested from limestone caves in Thailand and Malaysia (particularly the states of Sarawak and Sabah, on the island of Borneo), they were said to strengthen the lungs, enhance the complexion, and improve the overall health (sometimes referred to informally in terms of the immune system) of consumers. Prized for both their medicinal and status value, white, red, yellow, and black nests were also recommended for a wide range of ailments including weak blood, heatiness, cold, influenza, asthma, and convalescence after illness or surgery.[4]

Over a series of conversations, Hock Hua's Toa Payoh branch manager Tim explained that the red variety of bird's nest was particularly nourishing, but did not sell as well in Singapore because of the cost. Hock Hua carried white and yellow varieties of birds' nests in a range of grades (largely differentiated by the density of construction and size of pieces), typically displayed in a neat row of glass jars behind their consultation counter.[5] Although, by weight, birds' nests were among the most expensive items sold at Hock Hua (alongside cordyceps and high-grade ginseng), Tim reported that they constituted approximately 30% of the Toa Payoh branch's daily sales. During festivals like Chinese New Year, when items such as birds' nests were purchased for gifts and/or consumption at family meals, he noted, both overall and specialty item sales tended to increase.

At the time of my fieldwork, some Chinese medical physicians still practiced in "traditional" medical halls, perhaps assisted by a spouse or other family member. These consultation rooms were often located in small alcoves at the back of shops that sold both packaged products and bulk herbs stored in rows of small drawers behind the counter, carefully weighed and packaged in paper packets according to a physician's prescription. Most clinics and shops, however, had "upgraded" their environments and administrative procedures to some extent. Many physicians and Chinese medical institutions established orderly, efficient, and "modern" clinics with electronic queue boards in registration and physicians in white lab jackets. While some of these clinics still dispensed herbal packets, many had switched to liquid, pill, and/ or powder forms, thus accommodating Singaporeans' well-known preference for convenience within fast-paced city life. These developments, many of which were initiated by the Chinese medical community prior to TCM Practitioners Act (2000, revised and implemented in 2001), helped set the stage for further efforts to regulate the practice and products of Chinese medicine in Singapore.

The TCM Practitioners Act outlines the registration and professional guidance of Chinese medical physicians through the creation of the TCM Practitioners Board (TCMPB), one of five professional boards of the Ministry of Health.[6] The tasks of the TCMPB include approving, rejecting, revoking, or suspending registration applications, accrediting TCM courses and institutions, recommending continuing training for registered practitioners, regulating professional ethics and conduct, and otherwise fulfilling the provisions of the TCM Practitioners Act.[7] Couched in terms of safeguarding the public interest, this regulation differentiates legal from illegal practice by issuing or denying one-time registration certificates, and ensures ongoing appropriate behavior through annually renewable practicing certificates. The TCMPB also maintains a clear legal division between "TCM physicians" and "acupuncturists" with separate titles and registration categories, despite their common foundation in Chinese medical theory and practice.[8] Meanwhile, Chinese plant, mineral, or animal products are controlled by the Health Sciences Authority (HSA) – one of two statutory boards of the MOH, alongside the Health Promotions Board. Finally, Chinese *materia medica* are differentiated into two further categories: bulk materials, controlled under the Poisons Act (1970), and pre-packaged, over-the-counter products known as Chinese proprietary medicines (CPM), controlled under a variety of pieces of legislation and standards of practice maintained by the HSA.[9]

Insofar as it legitimized only specific educational institutions and individuals, this regulatory framework not only differentiated legal from illegal practice, but also sought to define authority and the proper transmission of knowledge and practice (both in terms of circulations between the PRC and Singapore, and within clinical settings). For instance, the TCMPB adopted the seventh edition of the Shanghai TCM University curriculum in developing Singapore's Chinese medical curriculum, to the exclusion of other Chinese medical theories or practices. Furthermore, based on the Singapore Medical Council's code of conduct, the TCMPB's Ethical Code and Ethical Guidelines for TCM Practitioners (released

January 2006 in English and Chinese) provides standards of clinical practice. The TCMPB's code guides physicians on the advertisement of services, information dissemination to the public, appropriate business and financial dealings, and proper relationships with patients and other physicians. With regards to clinical practice, it outlines modes of evaluation, delegation of responsibilities, duty of care, and medical record keeping.

In addition to these regulatory divisions (devised in relation to the biomedical model and biopolitics of the state), in the course of my observations at several large Chinese medical clinics in Singapore, I also noted disciplinary techniques, or anatomo-politics. Foucault describes the discipline of a manipulable, docile body politic – "subjected, used, transformed, and improved" – as a "microphysics of power" that begins with the distribution and ranking of people in space (Foucault 1995 [1975], 136). This process requires "enclosure" or "confinement," and individualized "partitioning" within sites such as factories, schools, hospitals, and military barracks. Next, the activities within these spaces are meticulously managed through an acute attention to time: idleness and wasted time is discouraged, a timetable is enforced and external rhythm imposed, precise sequences are defined, and bodily gestures prescribed (139–153). In sum, the productive capacity of a population is optimized through the distribution, enclosure, and partitioning of bodies-as-machines within institutions that enact biopower.

Reminiscent of Foucault's description of the hospital as a "curing machine," patients were processed within Chinese medical charity clinics with stereotypical Singaporean efficiency. Orderly spatial partitioning extended from arrival at the clinic to departure: first sorted as "new," "returning," or "regular," patients were numbered and divided by floor, by room, by physician, and then by bed. At each step, information was collected and transmitted to staff at the next step, "through systems of observation, notation, and record-taking which make it possible to fix the knowledge of different cases" (Foucault 1984, 287). Fine divisions and controls were thus implemented in the expedient processing of patients through these institutions, mirroring the ideological discipline of Singapore's "multiracial" population during post-colonial development.

Singapore's Chinese medical physicians moved within and between efficient healing environments that channeled flows of people, ideas, practices, and materials in accordance with state agendas seeking to maintain the economic and biological vitality of the population. Through ever-fine procedural prescriptions and proscriptions, the TCMPB-sanctioned curriculum and code of conduct dictated the manner in which physicians interacted with their patients and each other, employed certain tools (e.g., acupuncture needles) but not others (e.g., hypodermic needles), disseminated information to the public, took records, and shared knowledge. In a sense, even the circulation of *qi* and fluids within patients' bodies was regulated (indirectly) by the state via the designation of official forms of acupuncture, herbal preparations, and other modalities. Meanwhile, regional and international flows of Chinese *materia medica*, instruments, knowledge, and professionals – between the PRC or other Southeast Asian nation-states and Singapore, and then within Singaporean settings – were increasingly regulated by the

Singaporean government and mediated by biomedical/ scientific discourses of efficacy, safety, and quality control.

Circulation, convergence, and divergence

Laboratory-based studies and clinical trials of Chinese *materia medica* in the PRC, Singapore, the United States, and elsewhere have drawn Chinese medicine under the biomedical gaze, disaggregating and decontextualizing complex formulas and subjecting them to exogenous measures of efficacy and quality control. While biopharmaceutical researchers and professionals are thereby granted access to the "treasure-house" of Chinese medicine, Chinese medical physicians in Singapore are restricted to Chinese medical practices and products. In the event a Singaporean physician holds both biomedical and Chinese medical certificates, they are nonetheless required to maintain separation in their practices by operating two distinct clinics with separate entrances. Despite this mandatory separation, however, many aspects of biomedical anatomy, nosology, diagnostics, and treatment have been incorporated into Chinese medicine, particularly in the form of "TCM" (Hsu 1999; Taylor 2005). While the spatial and regulatory division of Chinese medicine from biomedicine in Singaporean biopolitics was explicitly upheld at the clinics in which I observed (by contrast with integrated clinics in the PRC), senior physicians flexed these divisions on a case-by-case basis (e.g., by incorporating biomedical diagnostics in their evaluations).

For instance, a Chinese medical physician with whom I worked, Dr Wang, received his earliest training in Singapore and then completed a Master's degree and three years of doctoral study at Nanjing University of TCM, where he also studied biomedical diagnostics and therapeutics. In Singapore, he maintained a private practice, volunteered at a large charity clinic one day a week, and volunteered his private clinic and time one day a week to a Buddhist charity organization. Until shortly after I left Singapore, the Buddhist charity also operated another clinic next door to his on the same day every week, where biomedical physicians offered free consultations. In Singapore, the physicians volunteering for these charitable clinics had to be clearly divided, but when conducting their annual medical mission (such as the one to Batam, Indonesia that I accompanied in March 2007), they were able to freely collaborate and share space. Although compliant with all legal requirements, in the course of his daily practice, Dr Wang often consulted x-rays and biomedical charts (which he either requested or were autonomously furnished by his patients). However, he could not fully apply the integrated training he had received in the PRC in his practice in Singapore.

On the other hand, the convergence of biomedical and Chinese medical practices in Singapore is in part facilitated by their frequent interaction – in the healthcare strategies of patients consuming both Chinese herbs and pharmaceuticals; in the cross-practice referrals between physicians; in the English-language acupuncture course offered to biomedical doctors and dentists at the Singapore College of TCM; and in the use of biomedical nosology or diagnostic techniques and equipment in Chinese medical clinics (blood pressure readers, x-rays, lab results, and so on). In the remainder of this section, I will highlight two further sites of convergence: namely, the linkages and disjunctures associated with issues of "lifestyle" in the first case, and with the safety and quality control of Chinese *materia medica* in the second.

In the course of establishing a healthcare system, the Singaporean government crafted an evolving notion of health that had to be communicated to the public. For this purpose, the Ministry of Health developed the Training and Health Education Branch in November 1963, which advocated awareness about nutrition, exercise, and general health maintenance. Under its current name, the Health Promotion Board (HPB) launched the National Health Campaign (or National Healthy Lifestyle Programme) in 1992. These efforts, alongside other public health campaigns, sought to reduce individuals' chances of contracting or developing the five major problematic diseases associated with rapid economic development: lung cancer, heart disease, hypertension, diabetes mellitus, and mental illness (Sinha 1995, 162). By the time of my research, stroke was added to the list, and the aims of the "healthy lifestyle" program were redesigned to promote regular exercise, healthy eating, abstinence from smoking, and stress management. Despite gains in sanitation, hygiene, and communicable disease control, the price of economic prosperity was reportedly being paid in the form of obesity, diabetes, and mental disorders.

Meanwhile, Chinese medicine also provided preventive or "healthy lifestyle" advice – an in-place, culturally embedded emphasis on diet, regular exercise, and mental calmness ("stress reduction" in HPB terms). In Chinese medical theory and practice, these concerns can be understood in relation to the concept of *yangsheng* ("nurturing life"), and its associated body of literature. In its broadest terms, *yangsheng* literature links the physiology of the body with the cosmos, encouraging a lifestyle that facilitates the harmonious resonance of cosmic, environmental, social/political, and physiological processes.[10] The techniques to accomplish this emphasize longevity over pathology and include dietary regulation, breath-cultivation, physical exercises, and "bedchamber arts" (Lo 2001, 22–27; Farquhar 2002, 261–267).

Chinese medical and biomedical professionals in Singapore thus share a concern about "diseases of modern life," although the methods and materials with which they classify, diagnose, and treat these issues differ in significant ways. Both recommend a moderate lifestyle, including proscriptions and prescriptions in types and quantities of food. To some extent, the moral authority of the state promotes an evolving set of personal and social responsibilities that guide the populace toward a lifestyle that privileges the health of the population (and thus the economy) over individual pleasures. On the other hand, there already exists a set of culturally appropriate practices guiding dietary regulation and other daily activities that, in my observation, sometimes carried more weight than the paternalistic admonishments of the state. This privileging of personal or cultural affinities over state agendas was particularly prevalent in Singaporeans' gustatory zeal.

In many respects, the Singaporeans with whom I interacted conformed to (and undoubtedly benefited from) the state's efficient, productive, and hygienic lifestyle – working long hours, maintaining a "healthy diet," exercising regularly, and so on. In other ways, they appeared to resist the state biopower agendas by smoking, idling, and drinking alcohol and continuing to eat rich, oily, and "unhealthy" foods in sense-provoking hawker centers (public eateries). Leisurely, messy, and communal meals thus acquired a new significance in the increasingly busy and sanitized cityscape in which Singaporeans eat. Older, outdoor hawker centers stood in contrast to air-conditioned restaurants and fast food establishments in shopping complexes, while Chinese feasts and medicines (even if they were essentialized constructions) were at odds with a decidedly non-Chinese "multiracial" national identity that privileges a biomedically defined "healthy lifestyle."

The value and meanings attached to Chinese food and medicine have also been problematized with respect to Singapore's role as a regional clearinghouse for herbs, food, and other products cultivated, produced, and/or processed in the PRC. As Yunxiang Yan (2012) observes, China's unmatched use of chemical fertilizers and pesticides, alongside large-scale adulteration and counterfeiting of food and medicine, has made food safety a particularly grave concern in the PRC and elsewhere. For instance, highly publicized cases of the production and distribution of poisonous food – tainted with substandard materials, banned additives, pesticides, toxic chemicals, or other materials – have catalyzed distrust, public outrage, and a series of national panics in the PRC. Such scandals, which involved the deliberate contamination of food, medicines, or animal feed for the sake of profit, were exacerbated by the apparent negligence or outright complicity of government agencies and officials. These practices have threatened people's health, social solidarity, and political stability in China, and prompted many countries to reject food and other goods directly exported from China.

The Singaporean government has addressed contamination and counterfeiting concerns on several fronts, seeking to secure the nation-state's role as a global trade hub. For instance, Singapore Customs' strict enforcement actions and regulations – such as the 2007 Secure Trade Partnership (a supply chain security program) – have sought to allay consumer fears that punctuate the flow of Chinese food and medicine. Hence, several of my respondents commented that they, and other Singaporeans, felt more comfortable buying Chinese food and herbal materials in Singapore than in the PRC (even if the products were cultivated in the PRC). Furthermore, under the Medicines Act (1975, revised 1977) and the Health Products Act (2007), good manufacturing practices (GMP) have been imposed on the manufacture and assemblage of CPMs in Singapore. Health Science Authority auditors thereby use GMP standards proposed by the World Health Organization (WHO) to dictate and ensure the consistent production and quality of CPMs.

Singapore has not been alone in its relatively recent scrutiny of "traditional" and "complementary" medicine (T&CM). The "WHO Traditional Medicine Strategy 2014–2023" reports considerable change in the global status and regulation of T&CM since the previous strategy was published in 2002. Between 1999 and 2012, 44 WHO member states developed T&CM policies, and 54 states began regulating herbal medicines; hence, by 2013, most WHO member states regulated herbal products to some extent. For instance, of the 37 states in the WHO Western

Pacific Region (which includes Singapore), six states drafted government documents or policies between 2000 and 2010, while six began regulating herbal medicine and nine implemented GMP standards between 2001 and 2010. In this region, Singapore and New Zealand are noted in particular for their "comprehensive regulatory frameworks for herbal and traditional medicines" (WHO 2013, 66). Meanwhile, in the WHO Southeast Asia Region, all member states except Timor-Leste had national T&CM policies in place by 2013, five states developed national T&CM policies between 2002 and 2012, and two states developed new regulations while three elaborated on extant regulations between 2003 and 2013. Despite progress in these regions and worldwide, the strategy reports, ensuring the safety, quality, and efficacy of herbal medicines remain challenges.

In addition to the regulatory and educational approaches promoted by the WHO, Singaporean researchers, physicians, and academics have addressed these challenges with techno-scientific solutions.[11] As Lee Tat-Leang (2006) explains, the practice of Chinese medicine and other so-called CAMs in biomedically-dominant societies raises a number of outstanding concerns. Beyond issues of professionalization and standardization (e.g., government oversight and the development of training institutions and curricula), he cites the need for improved communication between biomedical doctors and patients regarding CAM use; evidence of efficacy, safety, and cost-benefit analysis; medico-legal protections; and other efforts to integrate "conventional" and CAM approaches:

> Though it is widely perceived that "natural" products are safe, CAM use is not without risk. The co-use of prescription drugs and herbal medicines may cause both pharmacokinetic and pharmacodynamic drug-herb interactions. . . . The adverse effects of herbal medications may be unrelated to the herb itself, but may arise from manufacturing defects, e.g., misidentification, contamination, or adulteration. Thus, a pre-requisite is the authentication of herbs.
>
> (Lee 2006, 750)

While the practice of Chinese medicine and other CAMs presents a number of unresolved issues, Lee insists that one of the first steps to ensuring consumer safety is the authentication, or quality control, of herbs.

According to Singaporean researchers Kevin Yi-Lwern Yap et al. (2007), the traditional way of determining the quality (or authenticity) of Chinese herbs – based on age, origins, and physical characteristics – is not suitable to many of their contemporary commercial forms (capsules, powders, pre-packaged teas, and so on). Direct visual, olfactory, and/or tactile inspection of these products, they claim, cannot differentiate herbs like ground ginseng from adulterated or synthesized materials (Yap et al. 2007, 265–267). They therefore suggest that infrared spectroscopic "fingerprinting" and principle components analysis are more reliable means of authenticating Chinese *materia medica*, and thereby providing "quality assurance" (272).

Issues of authentication and quality control have also presented practical problems for Singaporean importers, wholesalers, and retailers. For instance, upon a brief return to Singapore in 2015, I had the opportunity to discuss the regional agarwood (*Aquilaria* genus; Mandarin: *chen xiang*) industry with enthusiasts and purveyors, as well as members of the Singapore Chinese Druggists Association. Indigenous to South Asia, then introduced to southern China and Southeast Asia, this fragrant, resinous heartwood and its derivative oil (often referred to as oud oil) has aesthetic, ritual, and medicinal uses, and is included in Chinese, Japanese, Ayurvedic, Tibetan, and Malaysian *materia medica*.[12] International trade of agarwood can be traced to at least the 13th century and expanded rapidly in the late 20th and early 21st centuries (major markets include United Arab Emirates, Saudi Arabia, China, and Japan), resulting in the addition of *Aquilaria malaccensis* to Appendix II of the Convention on International Trade in Endangered Species (CITES) in 1995. In the intervening decades, demand continued to grow worldwide and CITES designation was conferred on all other *Aquilaria* species, stimulating the establishment of government subsidized and private agarwood plantations in Southeast Asia.

The people with whom I spoke cited sustainability, cost, and authentication to be the most pressing issues surrounding agarwood – especially when it was sourced from intermediaries working with indigenous people in Sarawak or Sabah, Indonesia, Thailand, or other non-plantation locations.[13] According to members of the Singapore Chinese Druggist Association and others, by 2015, the market price for even the least expensive form of agarwood (namely, chips, which averaged over USD 6,000 per kilogram) had risen so much that most Chinese physicians and pharmacies in Singapore could not afford to prescribe or stock it. Even those who could afford to purchase it faced challenges differentiating genuine agarwood from fake specimens. In nearly every meeting I attended, I was shown an assortment of wood pieces (some appeared root- or branch-like, while others were chips or large chunks of wood) and was instructed to select the genuine pieces. With only minimal experience investigating agarwood, I found the task to be nearly impossible. While chemical composition analysis has isolated agarwood's distinguishing compounds (Chen et al. 2013), and gas chromatography-mass spectrometry has identified compounds in high-quality agarwood oils (Nor Azah et al. 2014), once again, such technology is not yet widely available.

In accordance with the values of a Singaporean Biopolis, a productive body politic, consumer safety, quality control, and secure supply chains have been prioritized in top-down regulation and enforcement on the one hand, and techno-scientific research and development on the other. Biopower is cultivated, in part, by means of the "healthy lifestyle" promoted by the state and in relation to a burgeoning transnational industry for T&CM products. The flow of Chinese food and *materia medica* – and particularly those originating in the PRC – is thereby marked by regulatory and scientific processes meant to allay consumer fears of contamination, and ensure the consistency, authenticity, and profitability of Chinese herbs, CPMs, and foodstuffs. Yet, techno-scientific solutions for these enduring issues are not uniformly or widely implemented, allowing "traditional" sensory methods of identification and grading to remain the most common methods of authentication and quality control. Governance, in this respect, seeks to act as a selective regulating matrix for the circulation of Chinese medicine and its convergence/ divergence with biomedicine.

Conclusion

The contemporary assemblage and governance of Chinese medicine vis-à-vis biomedicine in Singapore illustrates the fluidity and permeability of Chinese medicine. For much of the 20th century, Chinese medicine was regarded as "traditional" to China – a discrete category being defined within a bounded nation-state with a very different sociopolitical trajectory than Singapore's – and was therefore marginalized in favor of a biomedical healthcare system that more thoroughly represented the nationalist agenda. More recently, improved political and economic relations with the PRC and in the region (and increasing transnational circulations of people, ideas, and materials) – as well as a burgeoning global interest in T&CM – have destabilized these nationalist divisions. For instance, between 2011 and 2012, the global market for Chinese *materia medica* alone grew to an estimated USD 83.1 billion (an increase of over 20% in one year) (WHO 2013, 26). In light of its persistent popularity in Singapore and increasing economic potential elsewhere, Chinese medicine has been resituated in relation to both Singaporean biopower and the biomedical/scientific gaze in the 21st century.

With the increased circulation of Chinese *materia medica*, CPMs, and other T&CM medicine products, an increasing number of agents – policymakers, importers, researchers, transnational organizations, and others – press for the transparency and harmonization of its regulation. For instance, in addition to the WHO (2013) strategy discussed previously, Rob Verpoorte and an impressive array of contributing authors discuss the need for standards and research tools that ensure the safety, efficacy, and quality of "traditional" medicines (Verpoorte 2012). Fan et al. (2012) review regulatory frameworks of 11 countries/regions, with special emphasis on their approaches to safety, classification/definition, standards, and quality control. Meanwhile, Shaw et al. (2012) articulate a prevalent call for pharmacovigilance, arguing that post-market surveillance is necessary for herbal products that have not been pharmacologically or toxicologically tested, in the interests of public safety.[14]

The conjoined problems of status and authority – and the manner in which Chinese medical physicians negotiated their practice with respect to biomedical standards and values – are recurring themes I encountered in Singapore. In particular, issues of safety, authenticity, consistency, and quality framed Chinese medical practice and materials in biomedical and biopolitical terms. China's most recent opening to the world, the potential healthcare savings of an in-place, culturally appropriate set of practices that "nurture life," and the value of Singapore's entrepôt role in burgeoning transnational T&CM markets further stimulated the repositioning of Chinese medicine. Meanwhile, Chinese medical physicians negotiated their status and the legitimacy of their profession with reference to professional experience, collective (historical) experience, the cultural-moral authority of "Asian values," and techno-scientific discourses. In light of these ever-fluid transformations and negotiations, overlapping concerns about the circulation and governance of Chinese medicine become particularly germane.

In conclusion, Chinese medicine is increasingly being redefined in response to transnational circulations of people, products, and knowledge, and in relation to regulating matrices that seek to dictate its convergence and/ or divergence with biomedicine. These transnational flows do not simply move Chinese medicine from one location to another, but instead transform its theory, practice, and authority. As I have explored with reference to Chinese medicine in Singapore, issues of circulation and governance intersect on many levels – from the embodied experiences and anatomo-politics of patients, to institutional practices and biopolitical agendas, to transnational market logistics. These mutually influencing and reinforcing processes emerge somewhat in tension, and somewhat in conjunction, with Euro-American forms of political economy and scientific epistemology. The implications of these negotiations and reformulations extend beyond problematizing authority and heritage in a singular sociopolitical milieu, suggesting Chinese medicine's transnational creative capacity within much broader fields of exchange.

Notes

1 Many contemporary scholars have challenged the reification of "traditional medicine" and the false dichotomy between "tradition" and "modernity" upon which it is based (see, for instance, Leslie 1976; Leslie & Young 1992). Similarly, historical and anthropological studies have called into question the "traditional" designation in the 20th- and 21st-century reinvention of classical Chinese medicine as "traditional Chinese medicine" (TCM). Nonetheless, the value-laden designators "traditional" and "modern" remain in development, public health, nationalist, scholarly, and popular discourses – sometimes leveraged for their associations with very different fields of authority, while other times applied uncritically or colloquially. I employ the terms as they appeared in the course of my fieldwork and the literature reviews that bookended it, using quotation marks to acknowledge their contestation while recognizing their persistence in emic discourses (for further discussion of this issue, see Smith 2018).

2 Singapore's split-stream education system was founded during the colonial period, when most ethnic Chinese children were taught in community-based Chinese-language schools, while the children of British expatriates and future civil servants were taught in English-language schools. In the mid-1960s, the post-colonial government instituted a bilingual public education system in which English was the language of instruction and "mother tongues" (Mandarin, Malay, and Tamil) were taught as mandatory second languages.

3 This majority has subsequently been sustained, producing a fairly stable 76–77% ethnic Chinese majority, as reported by the National Population and Talent Division (2014).

4 The Singlish terms 'heaty' and 'heatiness' were both used by the Singaporeans with whom I spoke and are likely derived from the translation of the Chinese term for 'heat' (*re*). While the specific bodily experiences described by these terms varied between people (and between episodes of heatiness), the most common symptoms included otherwise inexplicable headache, fatigue, lack of focus, giddiness, a sensation of bodily warmth, yellow build-up in the eye, mouth ulcers, sore throat, dry mouth and, of course, fever.

5 Tim also explained that Hock Hua did not stock the black variety of birds' nests because, although they were high grade, Singaporeans considered them dirty and therefore too time consuming to clean and prepare.

6 At the time of my research, the other four included the Singapore Medical Council, the Singapore Dental Council, the Singapore Nursing Board, and the Singapore Pharmacy Board.
7 Information on the TCMPB was collected from its website, www.tcmpb.gov.sg, and from an unofficial interview with the TCMPB Registrar in 2005.
8 In fact, most Chinese medical physicians with whom I spoke informed me they would never perform acupuncture without also sending a patient home with an herbal treatment/ prescription. Oddly, this artificial division also gave preferential treatment to biomedical doctors by allowing them to practice acupuncture – provided they had the proper certification – without violating the separation between "TCM" and biomedicine.
9 These include the Medicines Act 1975 (revised 1977); Medicines (Traditional Medicines, Homoeopathic Medicines and Other Substances) (Exemption) (Amendment) Order 1998; Medicines (Chinese Proprietary Medicines) (Exemption) Order 1998; Medicines (Labelling of Chinese Proprietary Medicines) Regulations 1998; Medicines (Licensing, Standard Provisions and Fees) (Amendment) Regulations 2003; Medicines (Prohibition of Sale and Supply) (Amendment) Order 1998; Medicines (Labelling of Chinese Proprietary Medicines) (Amendment) Regulations 2005; and Health Products Act 2007.
10 The literature in question dates back to the Mawangdui medical manuscripts, unearthed in the early 1970s from a burial mound closed in 168 BCE (Lo 2001, 19–20).
11 As Yan describes, Ulrich Beck's risk society theory suggests that manufactured risks like those associated with food-safety problems, or counterfeit medicines, are products of industrialized modernity. By contrast, Yan asserts that science, technology, and modernization are often regarded as the solution to food-safety problems and risks in China (Yan 2012, 720–722), an observation that resonates with my own research in Singapore.
12 *Chen xiang* is discussed in the *Mingyi beilu* (*Supplementary Records of Famous Physicians*) and is classified in Chinese medicine as an herb for regulating *qi*, particularly affecting the stomach, spleen and kidney channels.
13 One respondent informed me that indigenous people on the island of Borneo knew where agarwood trees had been felled and would dig up the remaining root structures for intermediaries in exchange for rice and other commodities. It was not clear, however, whether any of the parties involved in these transactions regarded this to be a fair exchange, or whether agarwood root has the same properties and value as the aboveground portions of the tree.
14 It is worthy to note that this argument disregards centuries of documented experimentation and use; Chinese and other "traditional" pharmacopeia, clinical notes, and case studies apparently do not furnish sufficient evidence to establish the pharmacological and toxicological profile of *materia medica*.

References

Anderson, Benedict. 1991. *Imagined Communities: Reflections on the Origin and Spread of Nationalism*. London: Verso.
Chen, Ching-Tong, Yu-Ting Yeh, David Chao, and Chung-Yi Chen. 2013. "Chemical Constituents from the Wood of *Aquilaria sinensis*." *Chemistry of Natural Compounds* 49 (1): 113–114.
Clancey, Gregory. 2012. "Intelligent Island to Biopolis: Smart Minds, Sick Bodies and Millennial Turns in Singapore." *Science, Technology and Society* 17 (1): 13–35.
Cochran, Sherman. 2006. *Chinese Medicine Men: Consumer Culture in China and Southeast Asia*. Cambridge, MA: Harvard University Press.

Croizier, Ralph. 1976. "The Ideology of Medical Revivalism in Modern China." In *Asian Medical Systems: A Comparative Study*, edited by Charles Leslie, 341–355. Berkeley: University of California Press.

Fan, Tai-Ping, Greer Deal, Hoi-Lun Koo, Daryl Rees, He Sun, Shaw Chen, Jin-Hui Dou, et al. 2012. "Future Development of Global Regulations of Chinese Herbal Products." *Journal of Ethnopharmacology* 140 (3): 568–586.

Farquhar, Judith. 2002. *Appetites: Food and Sex in Post-Socialist China*. Durham and London: Duke University Press.

Foucault, Michel. 1984. "The Politics of Health in the 18th Century." In *The Foucault Reader*, edited by Paul Rabinow, 273–289. London: Penguin Books.

Foucault, Michel. 1990 [1976]. *The History of Sexuality: An Introduction: Vol. 1 of the History of Sexuality*. New York: Vintage Books.

Foucault, Michel. 1991. "Governmentality." In *The Foucault Effect: Studies in Governmentality (With Two Lectures by and an Interview with Michel Foucault)*, edited by G. Burchell, C. Gordon, and P. Miller, 87–104. Chicago: University of Chicago Press.

Foucault, Michel. 1995 [1975]. *Discipline and Punish: The Birth of the Prison*. New York: Vintage Books.

Grice, Kevin, and D. Drakakas-Smith. 1985. "The Role of the State in Shaping Development: Two Decades of Growth in Singapore." *Transactions of the Institute of British Geographers* 10 (3): 347–359.

Hobsbawm, Eric. 1983. "Introduction: Inventing Traditions." In *The Invention of Tradition*, edited by Eric Hobsbawm and Terence Ranger, 1–14. Cambridge: Cambridge University Press.

Hsu, Elisabeth. 1999. *The Transmission of Chinese Medicine*. Cambridge: Cambridge University Press.

Hutchinson, John, and Anthony Smith. 1994. *Nationalism*. Oxford: Oxford University Press.

Kong, James Chin. 2003. "Multiple Identities among the Returned Overseas Chinese in Hong Kong." In *Chinese Migrants Abroad: Cultural, Educational, and Social Dimensions of the Chinese Diaspora*, edited by M. Charney, B. Yeoh, and C. K. Tong, 63–82. Singapore: Singapore University Press.

Lee, Tat-Leang. 2006. "Complementary and Alternative Medicine, and Traditional Chinese Medicine: Time for Critical Engagement." *Annals Academy of Medicine Singapore* 35 (11): 749–752.

Leh, Charles Mu. 1993. *A Guide to Birds' Nests Caves and Birds' Nests of Sarawak*. Kuching and Sarawak: Lee Ming Press.

Leslie, Charles (ed.). 1976. *Asian Medical Systems: A Comparative Study*. Berkeley: University of California Press.

Leslie, Charles & A. Young (eds.). 1992. *Paths to Asian Medical Knowledge*. Berkeley: University of California Press.

Lo, Vivienne. 2001. "The Influence of Nurturing Life Culture on the Development of Western Han Acumoxa Therapy." In *Innovation in Chinese Medicine*, edited by Elisabeth Hsu, 19–50. Cambridge: Cambridge University Press.

National Population and Talent Division (Prime Minister's Office), Singapore Department of Statistics, Ministry of Home Affairs, and Immigration and Checkpoints Authority. 2014. *2014 Population in Brief*. Singapore: National Population and Talent Division.

Nor Azah, M. A., N. Ismail, J. Mailina, M. N. Taib, M. H. F. Rahiman, and Z. Muhd Hfizi. 2014. "Chemometric Study of Selected Agarwood Oils by Gas Chromatography-Mass Spectrometry." *Journal of Tropical Forest Science* 26 (3): 382–388.

Ong, Aihwa. 2006. *Neoliberalism as Exception: Mutations in Citizenship and Sovereignty*. Durham and London: Duke University Press.

Porkert, Manfred. 1974. *The Theoretical Foundations of Chinese Medicine: Systems of Correspondence*. Cambridge, MA: The MIT Press.

Quah, Stella. 1989. "The Best Bargain: Medical Options in Singapore." In *The Triumph of Practicality: Tradition and Modernity in Health Care Utilization in Selected Asian Countries*, edited by Stella Quah, 122–159. Singapore: Institute of South East Asian Studies.

Scheid, Volker. 2002. *Chinese Medicine in Contemporary China: Plurality and Synthesis*. Durham and London: Duke University Press.

Shaw, Debbie, Graeme Ladds, Pierre Duez, Elizabeth Williamson, and Kelvin Chan. 2012. "Pharmacovigilance of Herbal Medicine." *Journal of Ethnopharmacology* 140 (3): 513–518.

Sinha, Vineeta. 1995. *Theorizing the Complex Singapore Health Scene: Reconceptualizing Medical Pluralism*, PhD Diss. Baltimore, MD: Johns Hopkins University.

Smith, Arielle. 2018. *Capturing Quicksilver: The Position, Power, and Plasticity of Chinese Medicine in Singapore*. New York and Oxford: Berghahn Books.

Taylor, Kim. 2005. *Chinese Medicine in Early Communist China, 1945–63: A Medicine of Revolution*. London and New York: Routledge Curzon.

Unschuld, Paul. 1985. *Medicine in China: A History of Ideas*. Berkeley: University of California Press.

Verpoorte, Rob. 2012. "Good Practices: The Basis for Evidence-Based Medicines." *Journal of Ethnopharmacology* 140 (3): 455–457.

Waldby, Catherine. 2009. "Singapore Biopolis: Bare Life in the City-State." *East Asian Science, Technology and Society: An International Journal* 3: 367–383.

World Health Organization (WHO). 2013. *WHO Traditional Medicine Strategy 2014–2023*. Geneva: World Health Organization.

Yan, Yunxiang. 2012. "Food Safety and Social Risk in Contemporary China." *The Journal of Asian Studies* 71 (3): 705–729.

Yap, Kevin Yi-Lwern, Sui Yung Chan, and Chu Sing Lim. 2007. "Authentication of Traditional Chinese Medicine Using Infrared Spectroscopy: Distinguishing between Ginseng and Its Morphological Fakes." *Journal of Biomedical Science* 14: 265–273.

5 "Health products" at the boundary between food and pharmaceuticals

The case of fish liver oil

Liz P.Y. Chee

The boundary between what counts as a "food" and a "drug" has always been subject to cultural determinants which vary with geography and time. Owsei Temkin has written of the "fluent relationship of food and drugs" in ancient Greece, for example, where Galenic medicine was a largely dietetic system (Temkin 2002; Grant 2000). In China, Ute Engelhardt notes that "no major distinction" between "the application of drugs and foodstuffs" was made prior to the Tang dynasty, when *materia medica* and *materia dietica* began to be separately codified (Engelhardt 2001).

For modern states, however, clearly demarcating foods from drugs became an important project (Pray 2003). With few exceptions, regulations were crafted with the assumption that every consumable substance could be located in one of the two categories, but not both. Indeed, the first such American law, the 1906 Pure Food and Drug Act (also called the Wiley Act) stipulated among other things that foods and drug should not be co-mingled, but remain "pure" in relation to each other. Yet this ideal of clarity has never been fully achieved. The emergence in the mid- to late 20th century of the category variously called "health products" or "dietary/nutritional/vitamin supplements" would create a significant gray zone, which has only expanded with time. Different countries have also come to treat this third category in different ways, some relating it more to food, some to drugs, or in the case of China, to both. In the United States, it is neither, the government giving the private sector full autonomy to shape and define "health products" outside almost all regulatory restriction (Institute of Medicine of the National Academies 2005). Among many consumers in both countries (and around the world), health products are now considered an indispensable third category of consumables beside food and drugs, and in extreme cases their virtual substitute. While the phenomenon is widely reported, analyzed, and sometimes decried, we know comparatively little about how it came to be.

By health products, I refer to those non-prescription pharmaceutical-like substances that are advertised as supporting general well-being, and thus do not require in most instances a doctor's prescription. Some "health products" originated as drugs before being re-categorized by states, either through regulation or because their producers (or users) were responding to regulations which imposed a financial cost. In other cases, they were more food-like to begin with and took on medical characteristics later. This category-shifting helps explain why such substances remain understudied in social studies of medicine and pharmaceuticals, or in food studies, although as boundary objects they can help illuminate the larger process of state (and cultural) regulation across both domains. Health products may have more easily entered global circulation than many foods and drugs precisely because of their ambiguous regulatory identity.

This chapter will contribute to a history of this phenomenon by focusing on an early example – fish liver oil – whose categorization remains ambiguous even in the 21st century. The product had its origins in Europe, but moved to the United States, and then China, allowing us to make some preliminary comparisons across cultures, but also illustrating the role of circulation and governance in shaping its identity. I will show how fish oil has been defined by its encounters with different governance regimes, based not so much on arguments about intrinsic qualities or even use, but fiscal and political priorities. The resulting rules or policies have had an effect on how the product has circulated, and have been influenced in turn by patterns of circulation.

Fish liver oil is an animal-based material, and animals even more than plants are commonly categorized in modern societies as food. Although food was also considered medicine in many pre-modern cultures (the acceptance of the concept of "nutrition" alongside the germ theory would reinforce this concept in modern times), the use of modern technology to process raw animal parts and tissues has also been instrumental in making them appear in some instances less food-like and more drug-like. As such, "health products" are also a topic of importance to the field of animal conservation and ethics, given that the medicalization of wild species like bears, tigers, and rhinos by producers has often contributed to their exploitation and/or endangerment (Gratzer 2005).

Fish liver oil was (and remains) one of the most prominent substances at the boundary between food and medicine. It is a commodity which originated in the West (as cod liver oil) but spread to East Asia (as shark liver oil), and is now commonly associated with both regions in global commerce. The complexity of its geographic circulation and bureaucratic categorization make it an interesting yet difficult substance to follow, despite – or perhaps because of – its eventual ubiquity. My cases will proceed in roughly chronological order and move geographically from West to East. They are not a comprehensive history of fish oil, but meant to suggest the range of permutations involving its governance, circulation, and categorization, and hence it's shifting location on the food/medicine boundary.

The American tariff

The oil from cod livers had long been used as folk remedy in Scandinavia, but only came into general use as a medicinal after migrating to Germany by the early 19th century. It began to be marketed in England and the United States as a "patent medicine" in the 1840s, though its medicinal properties were not universally accepted by physicians.[1] Fish was clearly a food in Euro-American cultures, but oil from fish was also commonly associated at the time with industrial uses (whale oil for lamps being the most prominent).

In 1873, the *British Medical Journal* reported that on both sides of the Atlantic the question of whether cod liver oil was a medicine or a food "is repeatedly arising, and receives opposite solutions." The principle forum in which this question arose was not yet medicine, but tariff policy. The journal reported that a New York-based importer had been taxed a duty of "960 dollars in gold" for 2,700 gallons of cod liver oil because it was deemed a "medicinal preparation." This was apparently the first time that cod liver oil had been singled out from the larger category "fish oil," which was subject to a much lower duty. The importer filed a lawsuit against the US government and won the case, but only after eight years of litigation. Even so, this did not result in a long-term regulatory stability in the status of cod liver oil or other oil-based products as either a food or drug ("Is Cod-Liver Oil Medicine or Food?" 1873, 764).

Tariff law was in a general sense following pharmacology, which in the same period was finding more and more medical uses for substances previously classed either as food or industrial materials. Oils existed at the very boundary of these categories, as many of them were digestible, and yet circulated for often non-digestible uses. In 1869, a long article in the *Journal of the Society of Arts* had expressed enthusiasm that "there are many fish oils that might yet be brought into use for commercial and medicinal purposes," and referenced research from as early as 1856 showing both shark liver and cod liver oils to be useful in treating ulcers. Still, the phrase "commercial and medicinal purposes" ignored the fact that tariff policies around the world were coming to treat drugs differently, and less favorably, than raw materials for domestic manufacturing, or food production. A substance that straddled bureaucratic boundaries, particularly if it involved revenue, was thus destined to be problematic (Simmonds 1869, 173).

Fish-based oils were not singular in this regard. Olive oil had originally been imported in the United States as a food, but by the 1910s its importers were worried over "a very largely increasing use of olive oil as a medicine" which threatened to place it in a higher dutiable category. Decisions made in laboratories, hospitals, pharmacies, and clinics thus had the potential to increase revenue for the government while imposing a tax on the materials' producers and importers, causing them in this instance at least to lobby Congress and "the American people" in order to buttress the association of olive oil as a food product and play down its medicinal qualities. As a result of lawsuits brought by olive oil importers, the US government promulgated rules in the 1910s classifying all ingestible oils, including fish oils, as food and thus "non-medicinal" (U.S. Congress 1913, 268). These categories were never secure, however, as governments continually gathered evidence of boundary-crossing uses, constantly challenging importers, manufacturers, and users regarding the "true" identity of such substances, and perhaps even influencing their ultimate use through the corresponding effect that taxes had on pricing, and hence availability.

American tariff disputes involving fish-based oils continued as late as 1947, when an American importer of shark liver oil filed for a refund on a USD 4 million duty. Under the then-existing tariff, shark liver oil was classed as an "advanced drug" and so dutiable at 10% of its value. The importers, however, claimed that it was a "crude drug," a category exempt from duty. Even here, fish oil was inherently ambiguous because it arrived at the border fully processed (like an "advanced" drug), but such processing had been crude or minimal in comparison to most substances classified as drugs. The 1947 complainants did not pursue the argument that shark liver oil was "food," however, as had the earlier olive oil importers, because the food category was now subject to the same duty as "advanced drugs" ("$4,000,000 at Stake in US Shark Oil Suit" 1947).

That the classification of a substance as a food or drug (or a "crude" or "advanced" one) could be based on or influenced by tariff law rather than laboratory-based science, or even clinical use, would have surprised many consumers, given the common sense that those categories are fixed by standards higher and more objective than the market and the revenue needs of treasuries. But the need for precision in classification would likely not have occurred unless significant amounts of revenue were at stake for both parties – the government and importers. In the American case, tariff policy would be the forum in which these definitions would be drawn, policed, and contested. In the United Kingdom, beginning in a slightly later period, national health insurance would be the arena of similar disputes involving the same product.

British health insurance

While tariff regulation tended to drive the debate over the proper classification of fish liver oil in the United States, insurance regulation played a similar role in the United Kingdom. As we have seen, starting in the mid-19th century on both sides of the Atlantic, pharmacologists were creating a number of useful compounds by mixing cod liver oil with other ingredients with confirmed "medicinal properties" like quinine, often as "proprietary formulas" trademarked by drug companies. By itself, the oil was often still denied the status of a drug because evidence of its effectiveness was mainly empiric, and it often entered the physicians' medicine cabinet as an ingredient in drug admixtures (*The Western Lancet* 1844–45, 217).

The ambiguity lessened in the early 1900s, when scientists and researchers confirmed that the human body required a group of yet unknown trace substances to stay healthy. These would later be called "vitamins," and cod liver oil was found to be particularly rich in two of them – A and D – which were deemed necessary for the prevention of night blindness and rickets. Thus did cod liver oil take on the new characteristic as a "nutritional supplement" for the delivery of vitamins, or as a 1931 article in the journal *Nature* described it, "a fine *medicinal oil* of high activity" ("Medicinal Cod Liver Oil" 1931, 538, emphasis added).

The discovery of vitamins not only helped establish "nutrition" as a branch of medicine (on both sides of the Atlantic), but also reformulated the definition of drugs to include substances meant to treat "nutritional deficiency," which was now considered a disease. Rickets and night blindness were "deficiency diseases." This helped elevate cod liver oil from "medicated" to "medicinal," and suggested it might even be a drug in its own right. It was largely to deal with the classificatory problem presented by nutrition, vitamins, and fish oil that a 1933 report by the Scottish Committee of the British Medical Association observed that "medicine is a progressive science, and a substance which had been considered as a food might presently have to be considered as a drug" ("Preparations Not Ordinarily Regarded as Drugs" 1933, 200; Frankenburg 2009).

This re-classification was complicated, however, by the institutionalization of national health insurance in the United Kingdom, beginning in 1912, including the establishment of a "drug fund" to which eligible patients could apply for reimbursement for drug purchases made on a physician's order. What was and was not a drug now required definition, as, like the tariff, large sums were now dependent on classification. British doctors complained that the drug fund often denied reimbursement for their prescriptions of proprietary medicines, such as those based on cod liver oil, because local authorities refused to classify them as drugs. "The insurance service was one with a limited income," to quote the Scottish Medical Secretary, and doctors were seen as too casual in prescribing proprietary medicines rather than "actual drugs." From the policymakers' point of view, it therefore made sense to place cod liver oil and other "borderline substances" outside of the category "drug" in order to protect the limited resources of the insurance fund ("Preparations Not Ordinarily Regarded as Drugs" 1933, 200).

Cod liver oil remained the major "drug," however, in the treatment of many deficiency diseases, and these were so commonly diagnosed that disputes were bound to continue. A 1934 memorandum of the Scottish Committee of the British Medical Association sought to bring order by stating that while "the Drug Fund provides for drugs and certain approved appliances, but not for foods . . . special machinery has recently been set up" to give decisions "where there may be some doubt as to whether any particular substance is, in fact, a food or a drug." Admitting the ambiguity, however, and dealing with it on a case-by-case basis, did little to stem the confusion. While the same memo advised physicians to prescribe "British pharmacopoeial preparations," it also recognized that "there is an undoubted tendency for doctors to prescribe proprietary medicines . . . (and thus) proprietary articles as such should not be disallowed" ("The Scottish Health Service" 1934, 13).

The issue was kept alive through the 1950s, not just because of differences of opinion between the British Medical Association (representing doctors) and the Health Ministry, but because drug funds were locally administered, allowing for a range of answers on the question of how drug-like or food-like cod liver oil actually was. In one report meant to clarify state policy, a Joint Subcommittee on the Definition of Drugs wrote: "Preparations whose primary purpose is to provide nourishment in established diseases shall be classed as drugs – e.g. protein hydrolysates, allergilac, cod-liver oil and malt."[2] Despite the clarity of this language, local authorities continued to sometimes treat cod liver oil as a food. In 1953, the British Medical Association seemed to backtrack in stating that the official reports "are no more than a guide, and that every case must be dealt with on its merit," although it strongly urged that all local authorities consider cod liver oil a drug.[3]

The classification issue in this instance seems to have been driven not only by national insurance, but its local administration. Behind this was also the lingering sense that cod liver oil, being mainly a vehicle for the delivery of vitamins, was more food-like. Food, after all, was also a delivery system for vitamins. In the absence of regulation, it would have made little difference to physician or patient, as long as the prescribed substance was effective. But as with the American tariff, crossing the line between food and medicine now become an action with potential financial implications for the state.

Food "fortification"

State governance of a food/drug boundary became even more complicated with the growing medicalization of the field of nutrition, public recognition of the idea of vitamins, and resulting social anxiety over "vitamin deficiencies." Fish oil, given the major role it was perceived to play in vitamin delivery, loomed large in all these developments.

The idea of "fortifying" foods with vitamins began when Harry Steenbock from the University of Wisconsin demonstrated in 1923 that certain foods were able to self-generate vitamin D after being exposed to ultraviolet light, a concept known as irradiation. Quaker Oats bought the patent to Steenbock's invention in order to irradiate its breakfast cereals, and soon "vitamin-fortified" products began to saturate the market. Fish oil began to lose ground to irradiation, because its taste and smell were difficult to mask. In 1932, however, researchers at Columbia University successfully created cod liver oil bread and cod liver oil milk while completely eliminating any fishy smell, thus masking the "medicinal element" from the senses while allowing it to be promoted in the products' labeling. Thus did cod liver oil begin to compete with irradiation as a means of fortifying vitamins in a range of food products.

The ambiguous (drug-like) category of fortified foods has been well-noted (Bishai and Nalubola 2002, 37–53). Rima Apple (1996) has described how that the idea of fortifying the body against ill health arose as a concern after the First World War and continued strongly during and after the Second World War, as part of a rising concern with physical fitness. The policies of states (in this instance to produce healthy soldiers) were thus not external to the phenomenon of food fortification, even if they created headaches for food and drug regulators. There was also increased suspicion among nutritionists and the public that advanced technologies of food processing were removing vitamins (which was actually provable in the case of such foods as white rice), thus requiring that food be "supplemented" in order to remain nutritious (Apple 1996, 13).

For consumers, the difference between a food and drug had partly to do with how it was administered, and even what it tasted like. As two members of the Departments of Household Arts and Dietetics at King's College wrote in 1939, regarding fish oils:

> Now these oils have the reputation of being difficult to take, and to most adults and many children are nauseating. This in adults is not surprising, for

they were brought up to look upon cod-liver oil as a medicine and not as a food. They were given huge doses of a tablespoonful or more three times a day. The properly brought up modern child, however, has a conditioned reflex for fish-liver oils – looks, indeed on fish-liver oils not only as a food but a pleasant food. They may actually be induced to eat an unusual and new dish by using a fish-liver oil as a condiment.

(Lindsay and Mottram 1939, 14–15)

The authors offered a number of recipes to "camouflage [fish oil] in the form of a sauce or mayonnaise," most involved heating the oil to help mask its taste (Lindsay and Mottram 1939, 14–15). Increasing the identification of fish oil with food, however, threatened its carefully won status as a medicine in the eyes of its manufacturers. Thus, the general manager of British Cod Liver Oil Producers (Hull) Ltd. responded to the nutritionists by arguing that his oil "is of such delicate character (presumably owing to the highly unsaturated nature of the fatty acids of the oil) that it is peculiarly susceptible to damage by oxidation, heat, and other adverse circumstances." He willingly recognized the "harsh unpleasant fishy character" of the oil, but offered alternatives for rendering it more palatable which did not involve making it a cooking ingredient (MacLennan 1939, 190–191). Preserving its administration as a supplement to food, and perhaps even its medicinal taste, were thus important factors in its marketing and identity.

Thus, even as cod liver oil was being mixed into foods and its medicinal taste disguised by nutritionists, pharmaceutical companies were looking to market other varieties of fish oil as ever more potent medicines. A 1932 article in *The Science News-Letter* declared that three scientists had found and proven the liver oil of halibut to be "more than a hundred times as potent as cod liver oil" ("Vitamin A Concentrated in Halibut Liver Oil" 1932, 227). Halibut oil was given the metonym "super-concentrated sunshine," to compete with the branding "bottled sunshine" associated with cod liver oil. In the United Kingdom, fish in the *scombridae* family were sold under the marketing label of "super-D oil," whose content of vitamin D, the manufacturer claimed, "is 500 times that usually found in cod-liver oil" ("Natural Liver Oil" 1936, 1162). Subsequent research by pharmaceutical companies eventually expanded the medicinal species of oil-producing fish to include herring and burbot. This was also an opportunity for fishing interests to turn what were formerly unpopular fish (the more common term now being "trash fish") into products of economic value. This was particularly the case with burbot, whose liver oil was marketed as "three to four times as potent in vitamin D . . . as good grades of cod-liver oil" (Fortier 1939, 215).

Shark liver oil as strategic (war) material

We have seen that the identity of fish oil as a food or drug shifted as it encountered tariff regulation, insurance-related classification, and marketing competition (from irradiation). In these forums the question of whether a substance was a food or drug was repeatedly raised – if never finally resolved – because of significant financial implications for states and/or producers. A fourth forum through which fish oil circulated with consequences for its identity was wartime production. While military regimes are famously attracted to classificatory schemes, their emphasis is on production for the sake of victory in war. War-time production policies thus often conflict with, ignore, or re-order, peacetime classificatory schemes established to regulate revenue or accomplish other goals.

Such was the situation when the United States mobilized for war, beginning in 1940, and found itself in need of providing military personnel and munitions workers a "nutritious" diet. Cod liver oil was considered essential by this time as a vitamin supplement. Nearly the entire American supply was imported from Norway, however, and hence cut off when that country was occupied by Germany in 1940. The entire product of the Norwegian fishing industry was re-diverted to the Axis (England 1929, 116–122). As global circulation of this and other goods gave way to wartime autarky, substitutes were sought.

Given this dependence on fish liver oil, but with the cod fisheries compromised by the war, the United States turned to a fish population closer to its shores – sharks – as an alternative source. In so doing, the United States would promote industrial-scale shark fishing and processing, not only in its own waters, but around the world. In 1927, the *New York Times* had described sharks as heretofore "useless industrially" in reporting early attempts to turn the animals into a resource ("All of Shark to be Utilized" 1927). In the 1940s, however, almost every part of the animal began to be used in the context of the war effort. A contemporary article in the journal *Science* gave a list of shark parts and tissues that were to be reformulated for human consumption. Skin, for instance, was to be processed into leather and then made into boots. Primarily, however, sharks would be hunted and caught for their livers. The same article reported a recent boom in shark fishing due to "the urgent demand for shark liver oil" that "is largely replacing Norwegian cod liver oil, now impossible to import." It concluded that "shark liver oil is now as valuable and as sorely needed as rubber or tin" (Duckworth 1942, 8).

Shark fishing flourished, especially off the coast of Florida, where it was sometimes called "vitamin fishing" (Dunlap 1942, 9–10). While journal articles were emphasizing the growing production of shark liver oil, the American public was sometimes made to believe they were still consuming oil from cod. A 1942 *New York Times* article showed a picture of a wartime lecture on dietetics in which cod liver oil was promoted (among other foods) as a rich source of vitamin A (Kaempffert 1942). The author claimed erroneously that cod liver imports had actually increased since the war. This suggests that manufacturers were wary that oil from sharks would be rejected by the public and preferred to disguise it as something more familiar.[4]

As in the present century, overfishing of sharks in the 1940s led to concerns about sustainability. At that time, however, it was commercial fishermen who raised the alarm that shark populations were facing depletion off American coasts.[5] They "call[ed] for restriction of the fishing season to the times when the liver oil has its highest natural vitamin potency."[6] It was not until the late 1940s, however, when vitamin A began to be produced synthetically, that shark fishing finally returned to sustainable levels, and most shark processing factories shut their doors ("Synthetic Vitamin A Halts Shark Industry in Florida" 1950).

One legacy of classifying shark liver oil as a "strategic material" was that debates over its status as food or drug were temporarily suspended, along with tariffs regulating its importation. It was not described as specifically a food or medicine, but a substance for the "enrichment" of artificially produced substitute materials like margarine, whose status as a food was also somewhat ambiguous. Its purpose was "to heighten vitamin-values." The heightened attention to vitamins as a result of the war made them emerge as a category separate in the public mind from either foods or medicines, a kind of energy-producing fuel for the healthy bodies of soldiers and workers with a charisma all its own. As Rima Apple has pointed out, "the Second World War proved a major catalyst in popularizing vitamins in American culture," and thus helping to pave the way for the post-war "health supplement" as a category literally supplemental to the now too-restrictive categories of food and drug (Apple 2001, 135–150).

"Health products" (*Baojianpin*) in modern China

The global shark fishing and processing industry was a legacy of the Second World War, as fishermen in Canada, New Zealand, Brazil, and Argentina joined those in California and Florida in hunting sharks to satisfy a now insatiable appetite among the Allied Powers. Although cod liver oil would return to the market in the late 1940s, shark liver oil would linger as a global commodity outside the United States. One Allied nation that would be particularly influenced by this wartime trade in the post-war years was China, which in the long run became one of the major producers of shark oil-based "health products" (Chee 2016).

European cod liver oil had established a market in pre-war China due to British drug companies importing the material through Hong Kong. Reducing the outflow of local currency was a concern of the post-war Republican government, however, and much Chinese (and Japanese) pharmaceutical research in this period was dedicated to finding substitutes for foreign drugs using local ingredients. Since all cod liver oil was then the product of Atlantic cod (Pacific cod being a different species, not then classified as medicinal, and mainly found off Alaska and Siberia), Chinese interest had early turned to sharks, which were abundant in central Pacific waters. Experiments to test whether the livers of sharks were as efficacious as those of cod took place in Chinese laboratories as early as 1933, when one Shandong University researcher produced a seven-page report validating shark liver oil's medicinal properties. He pointed out that even the Norwegians had begun hunting basking sharks, in addition to codfish, for their enormous livers. It was the turn to shark liver oil by the Allied powers in the 1940s, however, that would finally validate this early Chinese interest and set the stage for a post-war Chinese shark liver oil industry (Xie 1933, 279–285).

Before this period, sharks were caught in China mainly for their fins. Shark fins (but also sometimes bones) were well-established food ingredients in Chinese cooking. Their documentation as food (rather than drug) in the famous Ming compendium *Bencao Gangmu* shows that historically sharks were not considered to have medicinal properties, unlike other large carnivores such as tigers and bears (Chee 2016). Nor had the Chinese acquired a taste for shark meat, since, after finning, this was mainly used as fertilizer or animal feed. Shark livers were thus a byproduct of shark fishing, and normally went to waste.

As for cod liver oil, its status as a food or drug was of little concern to the Chinese state. From a regulatory standpoint, most drug-like products in Republican China, including fish liver oil, were governed under a singular type-category called *chengyao*, or patent drugs. Pharmaceutical companies during this period were only required to provide the Chinese names of ingredients used in their products. Even to this day, companies whose products are based on well-established formulas need not follow the strict screening processes applicable to companies producing "new drugs" (*xinyao*). While spot tests were also done to ensure that the drugs were safe to use, drug regulations during the Republican period were not stringent, and thus the porous border between a food and drug was maintained by custom and marketing rather than regulation (Li 2014, 36–37).

While a firmer set of drug guidelines was established during the early Communist period, the national priority was drug production rather than fine-grade regulation. This was especially so during the Great Leap Forward (1958–1961), when the emphasis on increasing production of medicines to "defeat Britain" (then still the top exporter of pharmaceuticals to Asia) became the slogan of a new chemistry-based Chinese pharmaceutical industry. One of these changes was the increased use of animal tissues in the production of medicines, either through farming of animals that were formerly wild, or more systematic hunting/fishing/collecting. Medicinal uses for many of these animals had historic sanction in Chinese medical texts or folk practices. In the instance of shark liver oil, however, the model was entirely Western, as sharks had little pre-existing identity within Chinese medicine, despite their importance within Chinese food culture as an ingredient in shark fin soup.

During the Great Leap Forward, shark liver oil emerged as a "satellite" project, a Chinese term for a product which would help "launch" the economy to a higher plane. In 1958, the Xiamen Fish Liver Oil Factory announced that it would surpass the English brand Scott's in both quality and quantity of creamy white fish liver oil over the next two years, but using sharks instead of codfish. There was presumably little of the Western consumer resistance to shark products in China, given the high cultural status afforded shark fin soup, unlike wartime America where shark products had no pre-existing consumer acceptance. Later that year, Xingsha (the new brand of Chinese shark liver oil, *sha* meaning shark, and *xing* meaning star) announced it had met its goal of overtaking Scott's a year and nine months ahead of schedule. The announcement claimed that "the new recipe not only replaced Scott's as the top consumer choice in the international market, thus bringing glory to China, and also helping the country save as much as three million yuan." Accompanying statistics showed a twenty-fold increase in Chinese fish oil exports in this period. Allowing for exaggeration in both the rhetoric and statistics, Xingsha did manage in this period to begin climbing into international markets, where it remains to this day as one of China's major producers of fish liver products (Gaijin Rubai Yuganyou Zhiliang 1958, 561).

In the period of Deng's reforms, the medicinal animal trade began looking to expand its markets beyond what was strictly regarded as "medicine" – substances which required the prescription or advice of a physician. Their solution was to market animal tissue under the new category of "health product" (*baojian pin*) – something between medicine and food – a strategy that had been proven in the Western world through sales of vitamins. Chinese health products, however, could also be marketed as the modern versions of an older Chinese category called "food cures" (*shiliao*), which basically meant foods that were considered health giving or health restoring. The term "food cure" is an ancient one, though one not unique to China.[7] In the 1980s, however, Chinese pharmaceutical companies sometimes grafted this older category of "food cure" onto the Western-influenced "health product" to give cultural depth to the more contemporary and scientific flavor of the latter term. This also helped to create ambiguity as to whether the substance was or was not a "drug."[8] This new version of *shiliao*, however, would mainly be sold as preventive rather than curative.

By the late 1970s, the People's Republic of China, like other modern nation-states, had established separate rules, certification systems, and bureaucracies to regulate foods and drugs despite (if not because of) historical ambiguities. This regulatory boundary was arguably even less flexible than that in many Western countries, such as the United States, in that substances marketed as "health products" still had to be officially registered as either a food (*baojian shipin*) or a drug (*baojian yaopin*), but not both. Even then, they had to meet certain criteria which the Ministry of Health clarified in the late 1980s. Accordingly, health supplements in the drug sub-category had to be derived from *zhongyao* or Chinese medicinals, be nourishing, protect health, and not cause long-term harm (Zhongyiyao Xinxi 1988).

Still, as in the United States and United Kingdom before it, there was confusion in China about where this boundary lay, given that the government still regulated food and drugs separately. The category health product remained powerfully ambiguous, to the profit of drug companies. Or to put it another way, its ability to cross the food-drug boundary, and provide both nutrition and health in one package, made it an ideal marketing category for the new, more open economy of the Deng period. The category was particularly important for producers (pharmaceutical manufacturers as well as farmers) of animal-based substances. Industrial farming had stabilized their supply, yet many of their products (such as bear bile and deer antler) were too ambiguous in their efficacy to qualify as official drugs. As ingredients within health products, however, their producers could increase their value, while finding a much wider and more regular market. The producers of fish liver oil, which was officially listed as a drug, sought to straddle both categories by manipulating dosage levels. The pure product was potent enough to qualify as a prescription medicine, but lower grades could be marketed directly to consumers in the less regulated health supplement category.

Although the health product branding was helpful to producers, it was not popular among Chinese regulators. Perceiving abuse by drug companies, the Chinese government warned in the late 1990s that they might scrap the category altogether, though it ultimately decided on a less drastic policy of tightening regulation by only removing the *baojian yaopin* (health product – drug) sub-category ("Jianzihao Yaopin Sannian Hou Jiang Xiaoshi" 2000, 28). In September 2001, for example, the Ministry of Health forbade sales of bear bile as a health product, emphasizing its status as a controlled drug.[9] Still, the term *baojian yaopin* was never fully retired; the China Health Care Association was still using it as late as 2011, while a paper published as recently as 2016 provided a market update on the category, suggesting that Chinese companies found it indispensable for branding purposes (www.chc.org.cn/; Ma 2016, 263–265). This suggests the relative ease with which Chinese drug companies are able to circumvent registration laws by manipulating nomenclature (Chee 2016).[10]

Conclusion

Using the circulation of fish liver oil through regulatory and commercial regimes in the United States, United Kingdom, and China, this chapter has shown that health products or supplements have emerged as a modern – yet ambiguous – category influenced directly or indirectly by state attempts to separate and regulate the categories of food and drug. In the end, it is political economy – and not laboratory or science-based classification – that has determined how these borderline substances are categorized and re-categorized. Parties manufacturing, importing, or otherwise dealing with these substances in the market became involved in their categorization in relation to state regulation. The need for a third category seems to have arisen everywhere that foods and drugs have been subject to differential regulation. This category has proven extremely profitable to companies who can locate their products there, given not only states' confusion as to how to regulate, but also the confusion of consumers as to where to draw lines between medicine and food.

Perhaps the ultimate surrender of regulatory authority over such substances was reached in the United States in 1994, when the passing of the Dietary Supplement Health and Education Act (DSHEA) by the US Congress, which required the FDA to officially recognize vitamins as a category separate from food, and thus not subject to regulation of any kind. By drastically reducing the regulatory power of the FDA, the vitamin industry has since grown into a multi-billion-dollar business. In recent years, the creation of subcategories – "nutraceutical," "functional food," and "dietary supplement," to give but three examples – is continuing to blur the distinctions between food and drug (Fortin 2009, 323). Although China still has regulations on the books, their spotty enforcement is a concession both to the ambiguity of certain products, and pressure from producers and distributers. The new neoliberal global trade environment is also one in which the weakening of rules in one country effects those in all the others, making it harder to regulate in a strictly national fashion.

Notes

1 John Bennett wrote in 1841 "I am not aware that it (cod liver oil) is prescribed to any extent in Great Britain," in contrast to its commonality in Germany (Bennett 1841, 1). The following year, however, it was being cautiously received in New York, at least by "extra-professional doctors" ("On the Medicinal Properties of Fish Liver Oil" 1842, 44).

2 The Definition of Drugs Joint Subcommittee of the Standing Medical, Pharmaceutical and General Practitioner Advisory Committee issued its first guidelines in 1950, and adjudicated a variety of disputes over "borderline substances" ("Determination of Whether a Substance Is a Drug" 1953, 169; "Prescribing of Cod-Liver Oil, and Malt and Glucose" 1953, 169; "Was it a Drug" 1955, 114–115).

3 The problem had still not been resolved by 1956, however, when one doctor wrote to the Association's journal demanding the "right to expect local medical committees and the referees to declare themselves and . . . know what the law is before being penalized for breaking it." (Lewis 1956, 11).

4 Even manufacturers had to overcome the stigma of using shark liver oil (*Alaska Life* 1945, 40).

5 The same article points out that part of the American harvest was shipped to Britain for margarine making and stock feed (a vitamin supplement for cattle) ("Save the Sharks! Strange Cry Raised on West Coast" 1942, 120).

6 Ibid. p. 120. It was generally believed that female sharks had the lowest vitamin potency during summer, which was the birth season.

7 The term *shiyi* (literally "food-medicine") was invented as early as the Zhou Dynasty. A list of foods that performed as medicine was compiled during this time under the title Shiwu Bencao. During the Tang dynasty, a more common term was *shizhi* or "food cures." The famous Chinese physician Sun Simiao was believed to have popularized the use of food as medicine and the term *shiliao* is generally attributed to him (Interview with Professor Zheng Hong on 1 June 2014).

8 In 1987, however, the Health Ministry released a list of food ingredients that could also be considered "drugs" (No. 8 of the 1987 Food Hygiene Law of the People's Republic of China [Zhonghua Renmin Gongheguo Shipin Weishengfa]). Some examples were Chinese rat snake, chrysanthemum, and gingko (Zhongyiyao Xinxi 1988).

9 See China.org.cn (http://finance.china.com.cn/stock/special/gzt/20120216/536711.shtml) Accessed 29 March 2018.

10 The China Food and Drug Administration (CFDA) website presently categorizes products according to the following categories: "food product," "health-food product" i.e. "health product" (there is no mention of the subcategory "health-drug product" or *baojianyaopin*), and "drug." See China Food and Drug Administration: www.sfda.gov.cn/WS01/CL0001/

References

"$4,000,000 at Stake in U.S. Shark Oil Suit." 1947. *New York Times*, May 18.

Alaska Life. 1945 (8): 40.

"All of Shark to be Utilized." 1927. *New York Times*, January 16.

Apple, Rima D., ed. 1996. *Vitamania: Vitamins in American Culture*. New Brunswick, NJ: Rutgers University Press.

Apple, Rima D. 2001. "Vitamins Win the War: Nutrition, Commerce, and Patriotism in the United States during the Second World War." In *Food, Science, Policy and Regulation in the Twentieth Century: International and Comparative Perspectives*, edited by David F. Smith and Jim Philips, 135–150. New York: Routledge.

Bennett, John Hughes. 1841. *Treatise on the Oleum Jecoris Aselli or Cod Liver Oil*. London: S. Highley.

Bishai, David. and Nalubola, Ritu. 2002. "The History of Food Fortification in the US: It's Relevance for Current Fortification Efforts in Developing Countries". *Economic Development and Cultural Change* 51: 37–53.

Chee, Liz P. Y. 2016. *Reformulations: How Pharmaceuticals and Animal-Based Drugs Changed Chinese Medicine, 1950–1990*, PhD Diss. Edinburgh University and National University of Singapore.

"Determination of Whether a Substance Is a Drug." 1953. *British Medical Journal* 4817: 169.

Duckworth, J. Herbert. 1942. "Shark Liver Oil." *Science* 96 (2485): 8.

Dunlap, W. A. 1942. "Vitamin Fishing in Florida Waters." *Domestic Commerce*: 9–10.

Engelhardt, Ute. 2001. "Dietetics in Tang China and the First Extant Works of *Materia Dietetica*." In *Innovation in Chinese Medicine*, edited by Elizabeth Hsu. Cambridge, UK: Cambridge University Press.

England, Joseph W. 1929. "The Norwegian Cod Liver Oil Industry." *Journal of the American Pharmaceutical Association* 18 (2): 116–122.

Fortier, E. J. 1939. "Vitamins from Our Own Fish." *Scientific American* 160 (4).

Fortin, Neal D. 2009. *Food Regulation: Law, Science, Policy, and Practice*. Hoboken, NJ: John Wiley and Sons.

Frankenburg, Frances Rachel. 2009. *Vitamin Discoveries and Disasters: History, Science, and Controversies*. Santa Barbara, CA: ABC Clio.

Gaijin Rubai Yuganyou Zhiliang. 1958. "Ganshang Yingguo Mingpai Sigetuo." *Yaoxue Tongbao* 6 (12).

Grant, Mark. 2000. *Galen on Food and Diet*. London: Routledge.

Gratzer, Walter. 2005. *Terrors of the Table: The Curious History of Nutrition*. Oxford: Oxford University Press.

Institute of Medicine of the National Academies. 2005. *Complementary and Alternative Medicine in the United States*. Washington: National Academies Press. www.ncbi.nlm.nih.gov/books/NBK83789/.

"Is Cod-Liver Oil Medicine or Food?" 1873. *British Medical Journal* 2 (678): 764.

"Jianzihao Yaopin Sannian Hou Jiang Xiaoshi." 2000. *Zhongguo Jingji Xinxi (China Economic Information)* 7: 28.

Kaempffert, Waldemar. 1942. "What We Know about Vitamins." *New York Times*, March 3.

Lewis, A. 1956. "Was It a Drug?" *British Medical Journal* 4958: 11.

Li, Yanming. 2014. "Ruhe Kexue Xuanze Yuganyou." *Zhongguo Baojian Shipin* 11: 36–37.

Lindsay, Jessie, and V. H. Mottram. 1939. "Vitamin D in Diet: Palatable Methods of Supply." *British Medical Journal* 1 (4070): 14–15.

Ma, Shuming, et al. 2016. "Zhongguo Baojian Yaopin Shichang Bingdu Yingxiao Chuanbo Yiyuan Yanjiu." *Jingying Guanlizhe* 16: 263–265.

MacLennan, K. 1939. "Vitamin D in Diet." *British Medical Journal* 1 (4073): 190–191.

"Medicinal Cod Liver Oil." 1931. *Nature* 127 (3205): 38–539.

"Natural Liver Oil." 1936. *British Medical Journal* 1 (3935): 1162.

"On the Medicinal Properties of Fish Liver Oil." 1842. *New York Lancet* 2: 44.

Pray, W. Steven. 2003. *A History of Nonprescription Product Regulation*. Binghamton, NY: The Haworth Press.

"Preparations Not Ordinarily Regarded as Drugs." 1933. *British Medical Journal* 2 (3797): 200.

"Prescribing of Cod-Liver Oil, and Malt and Glucose." 1953. *British Medical Journal* 2 (4817): 169.

"Save the Sharks! Strange Cry Raised on West Coast." 1942. *Science News-Letter* 41 (8): 120.

Simmonds, P. L. 1869. "On the Useful Application of Waste Products and Undeveloped Substances." *Journal of the Society of Arts* 17 (846): 169–186.

"Synthetic Vitamin A Halts Shark Industry in Florida." 1950. *New York Times*, July 23.

Temkin, Owsei. 2002. *On Second Thought and Other Essays in the History of Medicine and Science.* Baltimore: Johns Hopkins University Press.

"The Scottish Health Service." 1934. *British Medical Journal* 2 (3835): 1–18.

U.S. Congress. 1913. "Timely Talks on the Tariff." *Congressional Serial Set.*

"Vitamin A Concentrated in Halibut Liver Oil." 1932. *The Science News-Letter* 21 (574): 227.

"Was it a Drug?" 1955. *British Medical Journal* 2 (4948): 114–115.

"Weishengbu Qiangdiao: Zhongyao Baojianyao Yilü Buzhun Baoxiao." 1988. *Zhongyiyao Xinxi* 2.

The Western Lancet (Cincinnati). 1844–45. 3: 217.

Xie, Ruli. 1933. "Sha Yuganyou Zhi Chubu Yanjiu." *Shandong Daxue Kexue Congkan* 1 (2): 279–285.

6 Globalized planta medica and processes of drug validation

The Artemisinin Enterprise

Caroline Meier zu Biesen

In recent years, access to artemisinin-based combination therapies (hereafter ACTs) as the first-line treatment for severe malaria has expanded substantially. In 2016, about 400 million ACT treatment courses were delivered to both the public and the private sectors in malaria-endemic countries (WHO 2016). In order to produce these quantities, immense volumes of leaves of the Chinese plant *Artemisia annua L.* (hereafter *Artemisia*) are required. The planta medica is today a key player in global efforts to help control malaria (Dalrymple 2012, 57). While for a long time, commodity production was confined mainly to Southeast Asia, Novartis – the leading pharmaceutical company in ACT production – has established an industrial partnership with the German-British-Swiss consortium Botanical Extracts Limited (BEEPZ) in East Africa. With financial funding of more than USD 25 million from Novartis and amounts in the millions from the Aga Khan Foundation, the UK Department for International Development (DFID), and the New York Acumen Fund, the private enterprise BEEPZ could build a modern production facility to extract the "white gold" (Bate 2008, 6).

Since 2007, the extraction facility built in the Kenyan Athi River industrial region has been producing artemisinin using recrystallization from the vaporized *Artemisia* raw extract – a viscous oil – and processing it further into artemisinin crystals. The crystals are distributed to Novartis, combined with lumefantrin in China, India, and the United States, and find their way back into developing countries as the combination drug Coartem®.

The commercial manager of BEEPZ, whom I shall call Dr Kane, credits the current anti-malaria policy of the World Health Organization (WHO) as foundational. "We produce artemisinin for the only pharmaceutical company currently able to supply malaria treatments to the WHO and the Global Fund. We are proud to state that the bulk of our production goes to provision of life-saving therapies sold at low cost to the poor," he told me. "Our goal is to develop a reliable, ethical raw material supply chain to ensure high-quality artemisinin returns."

East African farmers pioneered the commercial cultivation of high-yielding *Artemisia* hybrids (Ellman and Bartlett 2010). BEEPZ and co-financers of this multimillion-dollar project promised Tanzanian farmers that they would earn four times as much from a harvest of *Artemisia* as they would from ordinary cash crops. In practice however, the production-marketing chain was affected by high price fluctuation of artemisinin – sufferers being mainly small-scale farmers, who received low prices or experienced total loss of income (Dalrymple 2012).

A pharmacologist, whom I call Dr Bashiri, invested more than 15 years in the development of artemisinin production in East Africa. He confirmed that global markets are incorporated via medical commodities such as artemisinin. "Artemisinin is a daunting commercial product, a scary commodity," he stated. "Artemisinin prices rose globally, and then fell back again, which is hard from a producer's point of view." Dr Bashiri has developed a keen understanding of the market maneuvers, as well as the artemisinin boom-and-bust scenario, which had a great negative impact on both producers and health outcomes. However, he dismissed criticism that BEEPZ's method of producing *Artemisia* is exploitative: "We are under global pressure to extract more artemisinin, but the price we pay to farmers is fair." He regrets Tanzania's limited efforts to integrate itself fully into the global market of ACT production: "If we could manufacture ACTs in East Africa, we could compete with Novartis and become independent. But global players came from outside and the [Tanzanian] government's agreement to take the drug Coartem® [as first-line drug] hindered our enterprise." Interestingly, Dr Bashiri does not speak of gaining financial profit in this regard. "There is profit," he stated, "but what Novartis reaps is social profit. They move in the right circles with the WHO and are considered as the world saviours of malaria."

The developments I witnessed surrounding local *Artemisia* production – the euphoric mood at the outset, together with the disappointments experienced in particular by small-scale farmers over the years – have been subject to lively discussion in Tanzania. Since 2006, I have been investigating the consequences of the global transfer of *Artemisia* and the corresponding therapeutic and socio-cultural practice in Tanzania, taking into account the influence of global market and control mechanisms in the area of healthcare provision. This has enabled me to comprehend the complex landscape of actors involved in the commercial marketing of the planta medica.[1] For many, the story of the triumph of *Artemisia* as a medicine soon transmuted into a lesson about false expectations, the unpredictable and non-transparent dynamics of the global market, and the ruthless enforcement of profit-led interests by multinational companies (see Bate 2008).

With an eye on the marketing of Western pharmaceuticals, Van der Geest et al. (1996) speak of a dialectic tension: pharmaceuticals can lead to greater autonomy in health services, but also to new – in particular, economic – dependencies. Current medical-anthropological studies voice a growing concern about the humanitarian consequences of a "pharmaceutical expansion" (Greene 2011; Pollock 2011). Analyses of the role of pharmaceutical companies and Western pharmaceuticals discern two levels of inquiry. The first investigates cultural, scientific, and economic practices that have led to the powerful influence of the pharmaceutical industry. The second level seeks to understand how this influence (including the circulation of medicines) has grown and how it impacts on health practices at a local level. The microanalytical method developed by Kleinman (1992) has been incorporated into the concept of the "pharmaceutical nexus" (Petryna and Kleinman 2006), with a view to sharpening our vision for the way in which *local moral worlds* (i.e. social processes surrounding illness, neediness, and suffering) are administered, steered, and determined, as well as which (new) political functions global pharmaceuticals fulfill in these processes. This concept attempts to provide a résumé of a broad set of political and social transitions taking place within the global circulation of pharmaceuticals and to correlate these with one another. It is worth noting that theory of the pharmaceutical nexus has provoked analyses in the field that have widened our research focus; the activities of governmental and non-governmental organizations, for instance, have been subjected to inquiry in relation to national and global legislation (e.g. trade-related aspects of intellectual property rights [TRIPS]). Thus, it is possible to demonstrate how decisions made at a global level influence (both positively and negatively) the health of local populations. The standardization of medicines appears, on the one hand, as a necessary condition for biomedical methods of production. However, on the other, it facilitates the development of monopolies over strategic knowledge.

Central to the governance surrounding the commercialization and circulation of *Artemisia* are knowledge complexes attending the strategies deployed to market and distribute the plant. As I will demonstrate in this chapter using the pharmaceutical nexus, powerful actors were involved in the process of knowledge acquisition, as well as the pharmacological development of the artemisinin agent, the high-yield hybrid form, and the patenting procedure. Global circulation of the commodity derived from *Artemisia* – artemisinin – is in turn determined by dyadic alliances between these actors, such as the WHO, the pharmaceutical industry, philanthrocapitalistic organizations, global donors, and private enterprises, as well as global and national policymakers. By observing the Artemisinin Enterprise market construct through the lens of the (renewed) links to global public health policies driven forward by the WHO and nation-states (here: Tanzania), I am in a position to examine in this chapter the multiple entanglements between governance and circulation. The leading idea is that actors operate at various levels, have unequal control and access with relation to drug flows, and they seek to pursue their personal or group interests and thus steer the processes of production, marketing, dispensing, and use of artemisinin-based pharmaceuticals. I argue that these can be fruitfully viewed as multilevel processes – "level" here refers to the international, national, regional, and local tiers of social organization (see Van der Geest 2011) – in which political power, commercial power, and individual agency play prominent roles. In particular, I pursue the question of how circulation and governance are mutually dependent, and the nature of this specific and highly complex interactive relationship.

Ancient remedy, new promises: ACT innovation and governance

From plant to pharmaceutical

ACT innovation originally took place in China and continued in affiliation with and in cooperation between the WHO's Special Programme for Research and

Training in Tropical Diseases (TDR). The TDR was created in 1975 and was sponsored by the United Nations Development Programme, the World Bank, and the WHO. All the basic molecules assembled today in ACTs – whether artemisinin itself or its derivatives,[2] including the most frequently used ACT Coartem® – are the products of research programs undertaken by Chinese science and industry.

Viewed in retrospect, the "rediscovery" of *Artemisia* (Chinese: *qinghao*) – at least that relevant to the global context – was part of a historical process induced by the investigation of Chinese materia medica in the late 1960s (Yu and Zhong 2002). In 1967, in response to a request from the Vietnamese government for help on malaria treatment, the Chinese government launched a top-secret state mission – collectively referred to as "Project 523" or "Task-Force 523" – with the aim of finding a fast-acting drug to counter chloroquine-resistant malaria (see Su and Miller 2015). Military supervision highlighted the urgent nature of the research. Around 500 scientists investigating both known chemicals and traditional Chinese medicines worked on the extraction of components from Chinese herbal materials with possible antimalarial properties.

The turning point came when Professor Tu Youyou, pharmacologist and Task-Force 523 leader, found that an *Artemisia annua L.* extract showed a promising degree of inhibition against parasite growth. A document dated 8 March 1972 then bore witness to the scientific breakthrough: The active ingredient of *qinghao* was isolated as a colorless, needle-shaped crystal. In 1975, the name *qinghaosu* ("active principle of *qinghao*") was chosen for this new compound, later known in the "West" as the sesquiterpene peroxide "artemisinin" (Maude et al. 2010; Youyou 2011).[3]

The exciting results led to a nationwide effort to extract large quantities of the pure ingredient. Due to the prevailing political climate in China at the time, however, publications concerning artemisinin were restricted, with the exception of several published in Chinese. Moreover, during the Cultural Revolution, there were no practical ways to perform clinical trials on new drugs (Su and Miller 2015).

Chinese scientists[4] working on artemisinin then contacted the WHO's Steering Committee on Drugs for Malaria (CHEMAL), as part of the TDR, for pre-clinical laboratory testing of toxicity. They attempted in particular to obtain standard preparations of artemisinin (through toxicology and standard methods of production) in order to generate stable reference material. A series of presentations on artemisinin's antimalarial properties elicited an enthusiastic response at the Fourth World Symposium on Chemotherapy for Malaria, organized in 1981 by the WHO in Beijing. Despite knowledge of this antimalarial substance, it took a long time before decisions to deploy artemisinin on a global scale were made (Wright 2002). Bökemeier (2006, 97) describes the WHO bureaucracy surrounding the medicinal plant as a "chronicle of ignorance, misjudgement and procrastination."

Not until the mid-1990s did the application of artemisinin in Southeast Asia bolster hopes that a new antimalarial active ingredient had been discovered in the *Artemisia* plant. A significant body of trial evidence confirmed artemisinin drugs to be superior to any available alternatives (Bloland et al. 2000).[5]

The awarding of the 2015 Nobel Prize in Medicine to researchers identifying novel therapies marked an important milestone for research on infectious diseases. Professor Tu Youyou was awarded the prize for her key contributions to the discovery of artemisinin. Its impact on global health and the paradigm shifts in antimalarial drug research are the key factors that the Nobel Committee considered when they evaluated all nominations (Su and Miller 2015). The immense movement that took place after the artemisinin discovery required a full pipeline of research; moreover, a range of partnerships was initiated to build the research capacity needed in the countries mostly affected by malaria.[6]

A series of guidelines have been established to govern the transformation from plant to the frontline pharmaceutical. In the following section, I will discuss more closely the manner in which, on the basis of international binding regulations, the medicinal plant assumed the character of a commodity, and was thus able to circulate.

The golden standard: alliances promoting ACTs

A decisive precondition facilitating the circulation of the *Artemisia* plant and its partial synthetic derivatives was the affirmative stance of an extremely influential actor: the WHO. The WHO is in a position to exert influence over the circulation of drugs in a variety of respects. As a major global institution, the WHO aims to set binding (biomedical) standards in order to improve medical services for the world's population. Through close relationships with the member states, the WHO also exerts great influence over the guidelines devised by national governments.

As far as malaria is concerned, in 2000, the WHO launched the aid program Roll Back Malaria (RBM) under growing international pressure with the goal of reducing mortality by 2015. As a central aspect of the program, the WHO announced the application of artemisinin as a valid guideline: artemisinin (derivatives) were declared to be the most potent antimalarial active ingredients ever discovered. This WHO standard was akin to a "starting shot" for the race between international pharmaceutical companies for the active ingredient artemisinin (Bökemeier 2006, 3).

However, as the use of artemisinin-based monotherapies hastened the development of drug resistance, the WHO announced in 2006 the phasing out of monotherapies and its intention to focus instead on ACTs. From this point forward, not the acceptance of artemisinin but the validation of a specific drug made up a major aspect of the WHO's antimalarial program. The drug in question is based upon the derivative artemether, combined with lumefantrin, and marketed under the name of Coartem® (for application in developing countries) and Riamet® (for industrialized nations). As the preferred treatment method, this ACT earned much approval on a political level and advanced to the leading antimalarial therapy worldwide. Scientific studies on the drug's great compatibility buttressed this policy (Price 2013).

After a global consensus had emerged over delivering Coartem® to the neediest in resource-poor settings, the industry once again turned the requirements into market opportunities. A further influential actor comes into play for the circulation of Coartem®: the pharmaceutical company Novartis. In 2001, the WHO and Novartis pledged within the framework of a partnership alliance to produce Coartem® and to dispense it to developing countries at "cost price." Corresponding clinical testing of the application of the drug was conducted by the WHO, which is also responsible for marketing the drug globally. Novartis profits from this mediating role. Not only does the link to the WHO instill an "aura of trust" in the drug Novartis produces; the partnership also exerts a positive influence on the company's image to the outside world. Thus, Novartis no longer figures as a mere pharmaceutical company, but also as a philanthropic actor (for this perspective, see Ecks 2008, 16).

The opportunity to perform in the role of the savior of malaria-ravaged countries has bestowed high social profit on the producers of Coartem®. And this supposedly altruistic act has also paid for itself many times over: Novartis has been able to foster contacts to political decision-makers with considerable clout in key positions, and develop an (at least temporary) immunity against public admonishments of one-sided, profit-oriented company strategies.

The number of developing countries adopting Coartem® as the first-line treatment has since grown significantly, reaching 81 today (WHO 2016). The WHO recommendation for Coartem® was directly linked to an increased need for raw materials.[7] This not only necessitated expansion of the area under *Artemisia* cultivation; the goal was also to obtain a particularly high artemisinin content in the plants.

Artemisia seeds reached several European locations, notably a medicinal plants research institute in Switzerland (Médiplant) with close contacts to malaria control programs in East Africa. Research conducted over around 15 years by Médiplant on the biology of *Artemisia*, and the breeding on artemisinin permitted the development of cultivars with high artemisinin content, which facilitated the commercial production of *Artemisia* in the tropics (Wright 2002). Médiplant steers the marketing of this high-yield hybrid worldwide. Access to the plant and its commercial use is restricted via price and license rights: 10 grams of seed cost around USD 1,000 (see Von Freyhold 2008, 7). In addition to this, seed is only dispensed upon purchase of a costly license (USD 1 million) authorizing the purchaser to commercially market the raw material derived from the medicinal plant.[8]

The metamorphosis from plant to pharmaceutical also entails a structural transformation – in terms of its form –of the medicinal plant itself. As I will demonstrate in the following section, not only has a more potent plant emerged as a result of this circulation, but as a commercial commodity, *Artemisia* has also ended up in new sociocultural contexts in which it has co-determined commercialization (and consumption) processes. My ethnographic interviews with Tanzanian scientists involved in *Artemisia* research, ACT drug regulation, and development open perspectives on the bureaucratic and technological determinants of disease and health, and on the institutional ethics and medical and political practices that are determining disease management in the pharmaceutical nexus (see on this perspective, Petryna et al. 2006).

"The price of poverty": artemisinin-based pharmaceuticals in Tanzania

After years of new approaches and holistic innovations in malaria control, influential donors and major health agencies have returned to advocating top-down programs (cf. Nájera et al. 2011). As a result, the reliance on medical technology plays (again) a key role in the design of malaria-control programs (Cueto 2013, 30f). The WHO – together with key donor initiatives, such as the Global Fund (GFTAM), the World Bank Malaria Booster Programme (WBMBP), and the US President's Malaria Initiative (PMI) – claims that the problem of malaria could be solved by implementing well-designed magic bullets. Besides insecticide-treated mosquito nets, vaccines, and genetically altered mosquitos, access to ACTs is considered the most important factor in determining whether patients receive effective therapy (Meier zu Biesen 2018; WHO 2016).

These shifting global discourses on malaria control and funding patterns have had a decisive influence on how the Tanzanian government has rearticulated its malaria treatment policies. As shown by Kamat (2013), Tanzania has increasingly calibrated its national malaria control programs to meet with the previously mentioned global initiatives and (donor) funding opportunities. In doing so, the country has sought the more "reasonable" way forward in malaria control (i.e. the fundamental emphasis on technology-based interventions), rather than addressing concerns from community-based studies calling for more integrated approaches. A critical examination of the pharmaceutical nexus reveals that the emphasis on malaria control increasingly focuses on technical fixes, rather than on the social relations of inequality and poverty at the heart of the problem of malaria persistence in Tanzania and elsewhere (ibid., 213).

In 2006, the Tanzanian Health Ministry designated Coartem® – known locally as "ALU" (arthemeter/lumefantrin) or "mixed medicine" (Swahili: *dawa ya mseto*) – as the leading first-line antimalarial treatment. Novartis completed drug delivery in 2007 (Novartis 2010). As I engaged with the arguments advanced by Novartis' pharmaceutical executive and juxtaposed them with those of Tanzanian policymakers, I was able to sketch the logic underpinning the form of pharmaceutical governance represented in Tanzania's malaria policy. After the Tanzanian government addressed the needs of its population for ACTs, the global pharmaceutical industry's market options increased. However, the ability of Tanzanian actors to participate in this market was limited.

Between 2001 and 2007, a conglomerate of scientific institutions – including the Commission for Science and Technology (COSTECH), the University of Dar es Salaam, Tanzania Pharmaceutical Industries Limited (TPI), the National Institute for Medical Research (NIMR), the Institute for Traditional Medicine, the Government Chemist Laboratory Agency (GCLA), the Ministry of Industry, Trade and Marketing (MITM), and the Tanzanian Food and Drug Authority (TFDA) – elicited implementation strategies for the local production of ACTs. They were able to build on rich experience in the production of simpler artemisinin-based monotherapies. Novartis' early starting position, however, allowed it to determine the key factors for sole marketing.

In light of the developments in malaria control strategies, I interviewed the representatives of these institutions about the government's decision to roll out Coartem® throughout the country. Some spoke in positive terms about the policy, above all as it replaced the previous product sulfadoxine/pyrimethamine (SP), which had been associated with strong side effects, with a potentially more tolerable therapy. However, others expressed skepticism, mainly because the long-term sustainability of this bold initiative remained uncertain. The major criticism related to the preference for the drug Coartem®, which had not been convincingly substantiated in an objective manner. This particular drug had been incorporated into the WHO list of essential medicines, although other equally suitable drugs had been unable to make it past the complex pre-qualification procedure (Kachur et al. 2006). This fueled suspicions among the previously mentioned Tanzanian institutions and scientists that profit-seeking interests had exerted a hefty influence on the decision-making process. In their opinion, the Tanzanian drug industry's market was dismantled by both intellectual property law and the treatment policies which privileged proprietary transnational drug companies and the circulation of their drugs (for this perspective, see Peterson 2014). According to Tibandebage et al. (2016, 52f), the donors' large-scale tenders and the market entry requirement of the WHO pre-qualification shut out local Tanzanian firms from markets for human immunodeficiency virus (HIV), tuberculosis (TB), and malaria drugs, an effect that has been most damaging in antimalarials: While in 2006, about 90% of the then first-line treatment for malaria (SP) was sourced locally, Tanzania's shift to the more expensive first-line medication Coartem® excluded local firms. Two Tanzanian firms developed artemisinin-lumefantrin formulations but concluded that pre-qualification (costing an estimated USD 150,000) was unlikely to provide market access given the technology, scale and pricing power of Asian competitors.

> One has to view the whole thing as a process of globalization. We do not have strong pharmaceutical companies in Tanzania. And international companies have more power and money. Novartis was very lucky and was able to build a monopoly. . . . The Tanzanian government in turn collects taxes for imported medicines. Being poor is very expensive. In Tanzania there is no opportunity to produce combined therapies. That is the price of poverty.
>
> (Dr Kibahm, Director, Malaria Control Program, MOH)

> Artemisinin-based monotherapies had to [be] banned, Novartis created a monopoly and that has made us unemployed. If the WHO had waited, they would have been able to support local drug manufacturers. You wonder who benefits from this recommendation. People are poor and that costs us a lot. In the end, we depend on donors.
>
> (Dr Kalifa, Director, TFDA)

> We hoped that we would be able to start with the commercialization of *Artemisia* early on. The source of the raw material is guaranteed in our country. But now an expensive product is returning to us. If one wants to produce

ACTs cheaply, one needs a sense of economy. But the Tanzanians do not possess such a sense. And the government says that we cannot afford to produce ACTs by ourselves. Of course it is then easy for others to patent the drug.

(Dr Magoma, COSTECH)

As addressed in these quotes, the development of artemisinin drugs involves questions about intellectual property, resource distribution, and fair adjustment of profit. The configurations of capital and intellectual property (IP) have a structuring effect on both a drug's value and price. The processes of price determination and regulation by profits, which were excessively high for Novartis, illustrate the polarization of capitalism and contemporary pharmaceutical markets.[9]

The proprietary situation surrounding *Artemisia* is complex. Although China owns the intellectual property rights on artemisinin derivatives produced there (Shen et al. 2010, 119), it is possible to patent production processes as well as specific products. In the 1990s, Novartis and its Chinese partners signed a licensing and development agreement in which Chinese scientists – who had been experimenting with artemisinin combination therapies since the 1980s – developed the combination consisting of artemether and benflumentol (lumefantrin), the latter which they had also developed. Novartis formed a collaborative agreement with the Chinese Academy of Military Medical Sciences, Kunming Pharmaceutical Factory, and the China International Trust and Investment Corporation for further developing the same combination, eventually leading to international registration of Coartem® and Riamet®. Novartis supported its Chinese partners in building up local production opportunities and received in turn – after it had paid a fee of "several million dollars" (Dalrymple 2012, 12) – the rights to have the ACT patented outside of China (Maude et al. 2010). Through the patent, held in 49 countries (WHO 2010), Novartis has been able to secure itself a long-term monopoly. This monopolistic position is strengthened further by legally licensing the use of the hybrid seed. Access to the *Artemisia* variety suited to tropical conditions – and therefore of commercial interest – is restricted by a substantial financial hurdle.

As the WHO's decision to preference this drug has been the subject of much criticism in discussions I have had with national scientists, it was important to me to record the way WHO representatives in the Dar es Salaam branch of the organization saw the story. The WHO country manager welcomed the fact that the Tanzanian government had taken on the policy and, by virtue of this, its medical facilities had been equipped with the drug, especially with regards to Tanzania's limited technological capacities:

Coartem® is the best for the country. And that had to be implemented politically, although the Tanzanian government worried and said: "What should we do in the long term?" Yes, there were potential national producers of ACTs. In the meantime, WHO met with Novartis in Geneva. Indeed, the fact that we grow *Artemisia* in Tanzania and process the tablet in Europe is problematic. But do we really have the capital to make us independent in Tanzania? I don't think so. Certainly, one could reduce malaria and create jobs in our country, but we have no one who looks at this critically.

(Dr Kebede, WHO, Tanzania)

Contrary to the last statement, innumerable studies question the implementation of ACTs (in general) and the production cycle for Coartem® (in particular) (see in detail Meier zu Biesen 2013, 301f). Although Coartem® has been declared globally as the most sustainable malaria treatment, the conditions for its efficiency in Tanzania are not necessarily given. To begin with, the exceptionally long production cycle is associated with higher costs than was the case with previous products (Dalrymple 2012, 37f). The high cost maintains a dependence on multilateral subvention programs, thus detracting from more sustainable solutions. The vast majority of the implementation costs (96%) are covered by global donors. Without this subvention, Coartem® would remain unaffordable for the majority of the Tanzanian population (Mutabingwa 2005).[10] Second, timely access to authorized ACT providers (especially in rural Tanzania) is below 50% despite interventions intended to improve access such as social marketing and accreditation of private dispensing outlets (WHO 2010, 2; Khatib et al. 2013, 3). Third, the medical accountability at stake in this ACT policy has drastic implications for the Tanzanian population. Neoliberal health policies lead to an excess of counterfeits and illegal drug markets. The consequences of counterfeit medicines that do not meet the medical standards are devastating. Finally, despite the presence of low-cost, higher-quality therapies, the WHO recommended Coartem® as the preferred means. The fact that *this* ACT was validated by the WHO, but other – equally appropriate and cheaper – combination drugs did not succeed in the complex pre-qualification process leads to the suspicion that profit-oriented interests might have had strong influence on such decision-making processes.

There is no shortage of criticism leveled at WHO policy (Dalrymple 2012). Cheap medicine "gifted" by the pharmaceutical industry for developing countries should, according to the critics, be understood as part of a global price calculation strategy rather than a "perfect asset" (see Ecks 2008; Pollock 2011).

The WHO's response to this criticism was swift but measured (Kamat 2013, 194). It assured its critics that it was developing a new mechanism to facilitate access to ACTs alongside its global partners in the public and private sectors, especially the Global Fund. The multi-million-dollar Affordable Medicines Facility for Malaria (AMFm) initiative heavily subsidized ACTs. Reducing the retail price, it was argued, would ensure widespread distribution of ACTs, a concept developed further by the Roll Back Malaria Partnership (Gerrets 2010). Through a co-payment to certain accredited ACT manufacturers (here, Novartis), the program reduced the price of ACTs for first-line buyers such as governments by about 95%.

As shown by Petryna (2009), donors – not recipients – tend often to predominate, and the operations of international (health) organizations tend to reinforce existing and unequal power relations between countries. Tanzania provides an excellent case in point as public–private collaborative projects provide nearly all funding for ACTs (Kamat 2013, 211). The AMFm's capacity to meet its goals has been extensively debated, including how the structure of the distribution chain and nature of competition at all levels affects final prices (ibid., 255). Skeptics are concerned that the subsidy will be captured by middle-men within the private commercial supply chain and informal, unqualified, profit-maximizing retailers.

International aid organizations have been demanding that the WHO validate generic drugs demonstrating suitability as an alternative to Coartem®. Several public–private partnerships (PPP) are acting on this demand (Meier zu Biesen 2013). Following internationally voiced criticism, the demand to improve the poor access to ACTs through a forced participation of *national* actors in the production process was tabled (see Von Freyhold 2008). While Tanzanian actors, who produce and distribute part of the raw material, are (indirectly) involved in drug production, it is not possible to describe their involvement as a fair partnership model – as I will show in the next section.

Thriving in Tanzanian soil: *Artemisia* for the global market

Most of the commercial *Artemisia* plantations in East Africa are found in northern Tanzania, where smallholders cultivate the much sought-after raw material on around 4,000 hectares (TechnoServe 2007).[11] The cultivation of *Artemisia* (100% destined for export) is organized in a strictly hierarchical manner and follows WHO guidelines entitled "Good Agricultural and Collection Practices of *Artemisia*."

The private enterprise BEEPZ has developed one of the strongest technical bases for the production of artemisinin crystals. The main operating units of BEEPZ are the subsidiaries African Artemisia Limited Tanzania, East African Botanicals Kenya, and East African Botanicals Uganda. Moreover, non-governmental organizations (NGOs) act as mediators between farmers and BEEPZ. The NGO TechnoServe – which draws on funding from the WHO, USAID (the United States Agency for International Development), the BetterWorld Together Foundation, and private supporters – has partnered with BEEPZ to recruit local farmers.

Artemisia thrives in Tanzanian soil, which makes local production a viable commercial proposition for farmers. The viability of the Artemisinin Enterprise, however, depends on the costs and returns of *Artemisia* cultivation, which is a labor-intensive activity, especially at transplanting, harvesting, and threshing stages (Ellman and Bartlett 2010). Viability also depends critically on the current market price of artemisinin, which determines the price that processing factories can afford to pay farmers for dried leaves. Just as important as the actual price of artemisinin is the extent to which it fluctuates. Since 2004, there have been massive fluctuations, with prices varying between USD 1.6 per kilogram (when there was a world shortage on artemisinin following the WHO recommendation) to as low as USD 150 per kilogram at the end of 2007, when there was a glut (Hommel 2008). This see-saw effect, which is a direct result of the problems in forecasting demand, is not sustainable in the long term. Uncertainty over future demand, as I learned, has discouraged farmers from making long-term investments in production.

Farmers are paid according to a fixed set of regulations. Two factors determine the kilogram price of the raw material: weight and the artemisinin content. In 2007, the price calculation underwent modification, and from then on, it was the artemisinin content that largely determined the price. Prices sank when the content fell below 0.9% and rose proportionally with the measure of content (see Dalrymple 2012, 16). In 2013, the kilogram prices were once again standardized according to the artemisinin content. Farmers were initially encouraged by the prices that BEEPZ named; however, the reality is such that *Artemisia* cultivation has failed to turn a profit for many (see TechnoServe 2007; WHO 2010). In contrast to the dominant acquisition of the medicinal plant through international actors, farmers do not appear to participate in the profits commensurate with their involvement in production. Unrealistic prices of up to USD 3.30 per kilogram of dried leaves promised intermittently to farmers aroused unrealistic desires (Bate 2008). In relation to the actual end profit, the actual price paid to farmers – USD 0.40 per kilogram – means that the smallholders are not sufficiently compensated for their work. Furthermore, many producers have criticized the system of measuring the artemisinin content, to which the amount paid to farmers is coupled, as intransparent. Although new regulations – with bonus payments – is in planning, staff at TechnoServe are reserved when it comes to estimating the future development of farmers' earning potential:

R. SHAKWE (agricultural trainer, TechnoServe): Many farmers have not been paid for their leaves. They signed contracts, cultivated *Artemisia*, and when it was time to deliver, the prices fell. . . . Farmers were continually promised [improvement]. Some continued to grow *Artemisia* in the hope that the situation would improve, others gave up . . .

CM: [. . .] What kind of knowledge about the plant was communicated to farmers?

R. SHAKWE: We told the farmers that the plant was used to produce a medicine against malaria. The farmers asked whether they could use the plant themselves, but BEEPZ said we should tell the farmers that *Artemisia* is not suitable for personal use [in a medical-therapeutic sense].

Mr Mjatta was among the first farmers in Tanzania to start cultivating the plant. Unlike most smallholders, who farm *Artemisia* on average on just half a hectare of land, he presides over an 8-hectare plot boasting leafy well-tended *Artemisia* shrubs at the time of my visit. Mr Mjatta is employed as a commercial farmer for BEEPZ, and also functions as an agricultural advisor to smallholders in the area of *Artemisia* cultivation. According to him, farmers can profit from *Artemisia*, provided they have sufficient land and the suitable technology:

MJATTA: From a commercial perspective, *Artemisia* is a good cultivar. But yields are variable. This naturally has to do with a competition. Small-scale farmers are fearful of this competition. But malaria will increase, and that means continued need for medicines. I am also fearful that the prices will fall. BEEPZ could buy cheaper *Artemisia* leaves in China . . .

CM: How can the artemisinin content be more readily comprehensible for farmers?

MJATTA: That is a question of trust . . .

CM: . . . How is knowledge about the cultivation of *Artemisia* communicated to farmers?

MJATTA: USAID has paid a great deal of money to TechnoServe so that it imparts the knowledge [relating to cultivation] to farmers. But they get most of their knowledge from me.

CM: . . . Is the cultivation of *Artemisia* profitable for farmers?

MJATTA: At the moment, farmers are only being sold seedlings [not seeds]. Farmers are not allowed to take cuttings for themselves. Cultivation is attractive for farmers with large areas of land. Competition could increase the trading prices of artemisinin, but there is a monopoly. Only African Artemisia [Ltd.] buys the raw material and tests the artemisinin content, but we farmers are unable to account for this process. And on top of that, the [artemisinin] content depends on the weather and the soil. That is a trick. It was a motivational bait to hook the farmers at the outset and persuade them. They were lured by the premiums. Many farmers then lost their motivation . . .

Understood as both a method of critical analysis and inquiry, the pharmaceutical nexus is meant to capture a broad set of political and social transitions that fall under the globalization of pharmaceuticals – including the emic perspective in the observations about drug manufacturers. Pharmaceutical anthropology in this sense, as Van der Geest (2011) argues, should strive for a holistic insight into the entire culture of pharmacology, including much maligned area of manufacture. Encouraged by this investigative lens, I asked in interviews with representatives of BEEPZ how the company views the role and the position of its suppliers and local producers, and how the production of the combination therapy containing artemisinin is legitimized. The following statements of BEEPZ marketing managers reveal that manufacturers take a professional pride in putting a new – and better – drug on the market. Moreover, these statements disclose, what political drives are instantiated with globalizing pharmaceuticals and how local regimes of labor (here, Tanzanian farmers) affect such drives:

DR FOSSEY: One can make a lot of money from a medicine based on a natural active ingredient. . . . The challenge for us was to cultivate the material required cheaply. We are under pressure to manufacture the products cheaply and to retrieve more artemisinin content from a single plant. Our aim is to be able to produce the end product [ACT] here in Kenya. A good production chain, management and quality control are important. We believe that Africa can compete with Asian manufacturers. We are the chain, and we profit from this . . .

CM: How can farmers profit from this chain?

DR FOSSEY: We can safeguard their livelihoods. Each person in this production chain should receive a suitable income in order to guarantee the continuity of the chain. We support farmers. They have to understand that we are in a competition. If we fix the costs for the farmers too high, we cannot compete. It is also a social issue. Malaria is not just a business; it is also a matter of prestige, a social matter . . .

Another manager, Dr Rubanza, figures among the pioneers of artemisinin extraction and has accompanied the development of the Kenyan factory over many years from its inception. As I conducted the following interview, Dr Rubanza was no longer employed by BEEPZ and appeared – compared with other employees of the company – to speak a good deal more freely on the matter of production conditions:

DR RUBANZA: *Artemisia* excited us all. We traveled to Vietnam in order to learn the technology of extraction. Meanwhile, African Artemisia Limited had begun to focus in Eastern Africa on smallholders. Tanzania achieved the best result, they produced more than any other country. That is a question of strategy. . . . I have always asked myself why the actors [emphasis] *here* do not come together to harvest the knowledge and resources in order to produce the product locally. If TPI and BEEPZ would cooperate, then one could have agreed upon an ACT of our own. Now it has been agreed to take Coartem® and Novartis are the people producing it . . .

CM: . . . and profit from that . . .

DR RUBANZA: That is a tricky issue, the question of profit. What Novartis reaps is rather social profit. The actual price that BEEPZ gets for artemisinin is questionable . . .

CM: The price that the farmers get is questionable . . .

DR RUBANZA: No, no, no! The price at which we sold artemisinin to Novartis was fantastic. We are talking about USD 1,500 per kilogram. And the price fell, it is now perhaps only USD 300 . . .

CM: . . . and the farmers now get USD 0.4 per kilogram of *Artemisia* leaves . . .

DR RUBANZA: Yes, and that is a good price. But there is a danger that prices will sink further. The prices for the raw material are dictated by China and Vietnam. I know that the farmers work very hard. We had a good team of farmers, but we lost many of them. We were unable to pay thousands of farmers. But we want to retain them because they already have the knowledge.

CM: What kind of knowledge about *Artemisia* is communicated to the farmers?

DR RUBANZA: The farmers are given as much knowledge as is possible. We have training units, conduct field visits.

CM: Some farmers tell me that *Artemisia* is a poisonous substance, an insecticide, and that they were scared about touching the plant. Farmers reported that they had been informed by employees from the company BEEPZ that the plant is poisonous.

DR RUBANZA: We looked at the literature [as to] the effects of *Artemisia*, for example, during pregnancy. In the MSDS [Material Safety Data Sheet] there is information on artemisinin, but not on the plant. So, we tried to acquire knowledge of the plant ourselves and considered the types of danger that we might encounter. On this basis, we gave recommendations as to how this product should be treated during the harvest. Farmers were supposed to wear protective clothing. But on the other hand, there is no evidence that it is poisonous . . .

The way knowledge is hierarchized on local, national, and global levels reveals power structures. As demonstrated by the interviews, only a very limited knowledge of the *Artemisia* plant was communicated to farmers, and it is clear that in some cases false information was passed on purposefully.

The communication of knowledge here is primarily geared toward optimized harvesting techniques, and aims thus to generate the highest possible artemisinin content in the end product. As well as statements from farmers involved in production, their agricultural advisors have confirmed that the communication of knowledge of the raw material to be cultivated is controlled and selective. What is certain is that the farmers were influenced by the information they were given. This becomes clear, for example, with the fact that *Artemisia* (in its non-pharmaceutical form) was not consumed by producers as an antimalarial therapy. Control over the vegetative propagation and the explicit ban on growing cuttings from seedlings bought prevented, moreover, farmers from generating additional profits through private commercial activities.

Conclusion

Anthropological studies exploring the social and cultural dynamics of transnational therapeutic itineraries have shown that medical knowledge and practices become transformed as they circulate and experience global influences (cf. Pordié and Gaudillière 2014; Pordié 2008; Petryna 2009). Circulations, as these studies emphasize, lie at the core of recent attempts to explore the making of "global" knowledge and the encounters between socially and culturally distant worlds. A problematic issue with the circulation paradigm is that it tends to mask uneven and power-laden relationships (see Pordié 2015). This problem is, however, partially addressed in post-colonial science studies, which examine the relationship between techno-scientific knowledge and post-colonial orders. In this way, issues of mobility, asymmetries and power relations are revealed. As Dilger et al. (2012), for instance, emphasize, medicines, therapeutic practices, and healing practitioners – as well as institutions, technologies, ideas, policies, and the ethical frameworks to which they adhere – not only circulate, but also *shape* myriad aspects of social, political, and economic life. Moreover, when accounting for globalization, careful examination of mobility as an effect of power is required. This includes movements of health experts, WHO guidelines or policy documents across a wide range of social and cultural settings. Similarly, Lee and LiPuma's notion of "cultures of circulation" (2002) regards circulation as a generative process that entails negotiation at the nodes, as well as the performative channel making that enables flows, and hence creates spaces of new therapeutic opportunities and (new) forms of exclusion, control, and restrictions. Such an approach, I argue, allows to track things (here, medicines) through shifting social contexts, to elaborate hidden issues in the concept of circulation, its different levels, and its heterogeneity – and the way in which circulation and governance involve and reconfigure the objects in motion.

Artemisia as globalized planta medica provides valuable evidence of how traditional herbal formulations follow multidimensional itineraries across space and time – and as they travel through epistemological frameworks and local environments of use and practice, their attributes and their place in medicinal formulas are modified. In this chapter, I put particular emphasis on the conditions regulating *Artemisia* as it circulates across social space: the moments of its transformation, the contexts that determine its shape and those which repel it, the sites of synthesis with other forms of processes, and the sites of dislocation. The circulation of *Artemisia* acts both performatively – "extrovertly" – on the social contexts *Artemisia* circles through, but it also reconfigures the inherently mutable object of circulation, acting "introvertly." Circulation is therefore a central actor in global processes and understanding. Moreover, understanding governance as an arena of the multitude of actors and competing interests, within which many hubs for networking and negotiation have emerged (Kickbusch and Szabo 2014), allows investigating about the construction and hierarchization of knowledge, about ideological paradigms and the way they are created by influential institutions.

The medical rationale for governing and circulating artemisinin-based therapies, as I have shown, has been shaped by commercial demands in ways that have worked toward transforming the formerly scholarly Chinese medical tradition into a popular folk medicine (Hsu 2009). Every stage of development and production of the artemisinin drug Coartem® is characterized by a particular context, by the actors involved – together with their alliances – and by a discrete set of values serving to shape and motivate. By unpacking the organised sets of practices that govern artemisinin from the point of (re)discovery in China to global circulation within particular networks, I shed light on different forms of governance. In particular, I focused on the interactions between donors such as the WHO/Global Fund and China/Tanzania as an example of dealings between two different forms of governance: an emerging global (private) authority and an existing national government (see Szlezak 2012). Chinese scientists were the first to succeed in synthesizing artemisinin. This triggered a competition to produce the plant-based pharmaceutical drug artemisinin. The logic of need with regards to Coartem® follows political (and moral) estimations as to the potential threat of malaria. Novartis and the WHO formulate interest-based knowledge, which served as the basis for the validation of artemisinin. In so doing, the economic interests of the pharmaceutical industry are "re-declared" through WHO validation processes, which are based on the standards developed by these very same actors. The standardization of the ACT permitted the construction of a monopoly; the Tanzanian state has legislated for Coartem® as the treatment of choice, and has in doing so pushed the circulation of the drug forward decisively.

The logic by which *Artemisia* and artemisinin-based drugs circulate is shaped to a great degree by the pharmaceutical industry – which appears as the active center of the pharmaceutical nexus – as well as its relationships *to* and influence *on* other health policy institutions (via standardization, regulation, and forms of distribution). It is within this context that political demand for access to medicines is formulated. This touches upon a further conceptual aspect of the pharmaceutical nexus: the problem of the (in)compatibility of commercial, government, and non-government interests. As soon as medicines are produced, they are marketed as protected knowledge. Governments and pharmaceutical companies are involved in the production of this so-called "interest-based knowledge." I have examined the process by which this interest-based knowledge is governed using the example of Coartem®. The application of a multilevel perspective served as a tool to come to grips with these processes. A perspective that generally helps us to identify what is taking place at the various levels of pharmaceutical policy, governance, and practice; to provide explanations for the circulation of pharmaceuticals; and to point to the critical role pharmaceuticals have acquired in the dynamics of global health.

It appears at first glance that the transformations triggered by the *Artemisia* plant could lead to greater efficiency in the exploitation of *Artemisia* as traditional remedy: a "prescientific natural medicine" from traditional Chinese medicine is succeeded by "modern pharmacology," enabling us to extract the potential of this medicinal plant by combining modern science and highly developed technology. However, on closer investigation of the global circulation and governance of this medicinal plant, doubts about the foundation of this "linear concept of progress" emerge. This becomes apparent first and foremost on examination of the concepts of efficacy upon which the various actors base the marketing of artemisinin therapies. If we comprehend efficacy of a medicine as all factors upon which successful application depends (that is, its contribution to alleviating disease), then the pharmacological factor emerges as *one* – albeit central – aspect *among many*. A further factor with decisive influence on the long-term efficacy – and thus efficiency – of the medicine is the *actual* availability. However, for the Tanzanian reality (that is, for the health budget), the introduction of Coartem® initially meant a significant financial burden that the country was unable to stem on its own. Financial dependency on donors leads to a situation in which availability is dependent upon a range of imponderabilities, such as the future willingness and liquidity of those external donors (see Greene 2011). In this context, the priority of current malaria programs appears as a one-dimensional strategy in the face of a problem – malaria – requiring a multidimensional solution. An efficient malaria therapy can only then be developed when all important factors relating to efficacy are taken into consideration during the planning phase. This includes avoiding the shortage strategies arising from economic and political sectional interests in order to fully exploit the existing potential. The malaria-related pharmaceutical nexus – or the pharmaceutical modus operandi (Kamat 2013, 214) – provides an excellent example of how the global pharmaceutical industry has exerted significant influence on the neoliberal global discourses on malaria control.[12]

Notes

1 Through following its trajectory, meaning, and use, the *Artemisia* plant became my main object of research, and I have traced it to multiple locations where it is planted, used therapeutically, and commercially processed. Malaria in Tanzania became a

lens for exploring local and global health governance. Based on 14 months of ethnographic research undertaken from 2006–2008 and regular follow-up visits until 2015, I scrutinized why health levels have stagnated – or even declined – due to poverty, WHO malaria policy implemented by governments, and global power relations. I also reflected critically on the dialectics and reciprocities between different actors and their relationship to existing influential reference systems (such as the WHO and the pharmaceutical industry). In addition to field data, I used sources from the WHO Archive in Geneva, in particular those relating to the WHO's engagement with scientific artemisinin working groups in China.

2 Semisynthetic derivatives (such as artesunate, artemether, and arteether) are antimalarials derived from artemisin by chemical reaction.

3 For a detailed review of *Artemisia*'s history in TCM, see Hsu (2006). For a chronology of "Project 523," see Jianfang (2013).

4 Most of them were members of the Institute of Materia Medica, Chinese Academy of Sciences in Beijing, the Department of Pharmacology and Synthetic Chemistry at the Second Military College in Beijing, and the Institute of Parasitic Diseases, Chinese Academy of Sciences (see WHO Archive file 16/181/M2/61).

5 In Thailand and Vietnam, a number of studies conducted between 1987 and 1994 confirmed the Chinese findings of high efficacy and low toxicity of the artemisinins against malaria (Maude et al. 2010, 14). See also artemisinin research in collaboration with the WHO done by the "Institute of Natural Products Chemistry" in Vietnam. The institute is the national point of contact for the "Regional Network for the Chemistry of Natural Products" in Southeast Asia (WHO Archive file T 16/181/228).

6 Owing to the fact that many scientists were involved in "Project 523," the fact that the Nobel Prize was awarded to a single person has not been without controversy (see Su and Miller 2015).

7 Even though studies show a breakthrough in genetically decoding artemisinin (see Milhous and Weina 2010), the active pharmaceutical ingredient (API) or "raw artemisinin" is extracted exclusively from the *Artemisia* plant leaves.

8 Anonymous (personal communication, 14 June 2009).

9 See: "Between Financial Capitalism and Humanitarian Concerns: Value, Price and Profits of Hepatitis C Antivirals and Artemisinin Combination Therapies for Malaria." Paper presented by Maurice Cassier at the Workshop *The Making of Pharmaceutical Value: Drugs, Diseases and the Political Economies of Global Health* on 16 June 2016 in Paris (GLOBHEALTH project).

10 The cost price negotiated between the WHO and Novartis is USD 2.40. As a result of subventions, Coartem® is now available in public health institutions for around 50 cents (US).

11 For the production of 120 million ACT treatment units, roughly 114 tonnes of artemisinin are required. 10,000 square metres of land cultivated with *Artemisia* yield around 100 kilogrammes of pharmaceutically usable leaves, out of which one kilogramme of the active ingredient artemisinin can be extracted (Bökemeier 2006).

12 Acknowledgements: The editing of this chapter was supported by the ERC-funded project GLOBHEALTH, under European Research Council (grant number 340510). Editing was performed by Dr Gabrielle Robilliard-Witt.

References

Bate, Roger. 2008. "Local Pharmaceutical Production in Developing Countries: How Economic Protectionism Undermines Access to Quality Medicines." www.libinst.ch/pub likationen/LI-LocalPharmaceuticalProduction.pdf.

Bloland, Peter B., Mary Ettling, and Sylvia Meek. 2000. "Combination Therapy for Malaria in Africa: Hype or Hope?" *Bulletin of the World Health Organization* (78): 1378–1388.

Bökemeier, Rolf. 2006. "Ein Kraut wirkt Wunder." *GEO Magazin*. Leonardo da Vinci (6): 1–6.

Cueto, Marcos. 2013. "Malaria and the Global Health in the Twenty-First Century." In *When People Come First: Critical Studies in Global Health*, edited by J. Biehl and A. Petryna, 30–53. Princeton and Oxford: Princeton University Press.

Dalrymple, Dana G. 2012. *Artemisia Annua, Artemisinin, ACTs and Malaria Control in Africa: Tradition, Science and Public Policy*. Washington, DC: Politics & Prose Bookstore.

Dilger, Hansjorg, Abdoulaye Kane, and Stacey Langwick. 2012. *Medicine, Mobility, and Power in Global Africa: Transnational Health and Healing*. Bloomington: Indiana University Press.

Ecks, Stefan. 2008. "Global Pharmaceutical Markets and Corporate Citizenship: The Case of Novartis' Anti-Cancer Drug Glivec." *BioSocieties* 3 (2): 165–181.

Ellman, Antony, and Elspeth Bartlett. 2010. "Cultivation of Artemisia Annua in Africa and Asia." www.mmv.org/sites/default/files/uploads/docs/artemisinin/2010_Madagascar/Cultivation_of_Artemisia_in_Africa_and_Asia.pdf.

Gerrets, Rene. 2010. *Globalizing International Health: The Cultural Politics of 'Partnership' in Tanzanian Malaria Control*. New York: New York University.

Greene, Jeremy A. 2011. "Making Medicines Essential: The Emergent Centrality of Pharmaceuticals in Global Health." *BioSocieties* 6 (1): 10–33.

Hommel, Marcel. 2008. "The Future of Artemisinins: Natural, Synthetic or Recombinant?" *Journal of Biology* 7 (38): 1–5.

Hsu, Elisabeth. 2006. "Reflections on the 'Discovery' of the Antimalarial Qinghao." *British Journal of Clinical Pharmacology* 61 (6): 666–670.

Hsu, Elisabeth. 2009. "Chinese Propriety Medicines: An 'Alternative Modernity?' The Case of the Anti-Malarial Substance Artemisinin in East Africa." *Medical Anthropology* 28 (2): 111–140.

Jianfang, Zhang. 2013. *A Detailed Chronological Record of Project 523 and the Discovery and Development of Qinghaosu (Artemisinin)*. Houston: Strategic Book Publishing.

Kachur, Patrick S., Carolyn Black, Salim Abdulla, and Catherine Goodman. 2006. "Putting the Genie Back in the Bottle? Availability and Presentation of Oral Artemisinin Compounds at Retail pharmacies in Urban Dar es Salaam." *Malaria Journal* 5: 1–8.

Kamat, Vinay R. 2013. *Silent Violence: Global Health, Malaria, and Child Survival in Tanzania*. Tucson: The University of Arizona Press.

Khatib, Rashis A., Majige Selemani, Gumi A. Mrisho, Irene M. Masanja, Mbaraka Amuri, Mustafa H. Njozi, Dan Kajungu, Irene Kuepfer, Salim M. Abdulla, and Don de Savigny. 2013. "Access to Artemisinin-Based Anti-Malarial Treatment and Its Related Factors in Rural Tanzania." *Malaria Journal* 12 (155): 1–8.

Kickbusch, Ilona, and Martina Marianna Cassar Szabo. 2014. "A new governance space for health". *Global Health Action* 7: 23507, DOI: 10.3402/gha.v7.23507.

Kleinman, Arthur. 1992. "Local Moral Worlds of Suffering: An Interpersonal Focus for Ethnographies of Illness Experiences." *Qualitative Health Research* 2 (127): 1–9.

Lee, Benjamin, and Edward LiPuma. 2002. "Cultures of Circulation: The Imaginations of Modernity." *Public Culture* 14 (1): 191–213.

Maude, Richard J., Charles J. Woodrow, and Lisa J. White. 2010. "Artemisinin Antimalarials: Preserving the 'Magic Bullet'." *Drug Development Research* 71: 12–19.

Meier zu Biesen, Caroline. 2013. *Globale Epidemien – lokale Antworten: Eine Ethnographie der Heilpflanze Artemisia annua in Tansania*. Frankfurt and New York: Campus Verlag.

Meier zu Biesen, Caroline. 2018. "Artemisia Annua and Grassroots Responses to Health Crises in Rural Tanzania." In *Artemisia Annua: Prospects, Applications and Therapeutic Uses*, edited by T. Aftab, M. Naeem, and M. Masroor A. Khan, 17–40. Boca Raton, London, and New York: Taylor & Francis Group.

Milhous, Wilbur K., and Peter J. Weina. 2010. "The Botanical Solution for Malaria." *Science* 327 (5963): 279–280.

Mutabingwa, T. K. 2005. "Artemisinin-Based Combination Therapies (ACTs): Best Hope for Malaria Treatment But Inaccessible to the Needy!" *Acta Tropica* 95 (3): 305–315.

Nájera, José, Matiana González-Silva, and Pedro L. Alonso. 2011. "Some Lessons for the Future from the Global Malaria Eradication Programme (1955–1969)." *PLoS Medicine* 8 (1): 1–7.

Novartis. 2010. "Coartem in Africa: Gaining Momentum on the Ground." www.corpo ratecitizenship.novartis.com/patients/access-medicines/ac-cess-in-practice/coartem-in-africa.shtml.

Peterson, Kristin. 2014. *Speculative Markets: Drug Circuits and Derivative Life in Nigera*. Durham: Duke University Press.

Petryna, Adriana. 2009. *When Experiments Travel: Clinical Trials and the Global Search for Human Subjects*. Princeton: Princeton University Press.

Petryna, Adriana, Andrew Lakoff, and Arthur Kleinman. 2006. *Global Pharmaceuticals: Ethics, Markets, Practices*. Durham: Duke University Press.

Petryna, Adriana, and Arthur Kleinman. 2006. "The Pharmaceutical Nexus." In *Global Pharmaceuticals: Ethics, Markets, Practices*, edited by Adriana Petryna, Andrew Lakoff, and Arthur Kleinman, 1–32. Durham: Duke University Press.

Pollock, Anne. 2011. "Transforming the Critique of Big Pharma." *BioSocieties* 6 (1): 106–118.

Pordié, Laurent. 2008. "Tibetan Medicine Today. Neo-Traditionalism as an Analytical Lens and a Political Tool." Introduction in *Tibetan Medicine in the Contemporary World: Global Politics of Medical Knowledge and Practice*. London and New York: Routledge.

Pordié, Laurent. 2015. "Previous Drugs. Making the Pharmaceutical Object in Techno-Ayurveda." *Asian Medicine* 9 (1): 1–28.

Pordié, Laurent, and Jean-Paul Gaudillière. 2014. "The Reformulation Regime in Drug Discovery: Revisiting Polyherbals and Property Rights in the Ayurvedic Industry." *East Asian Science, Technology and Society: An International Journal* (8): 57–79.

Price, Ric N. 2013. "Potential of Artemisinin-Based Combination Therapies to Block Malaria Transmission." *Journal of Infectious Diseases*, Editorial Commentary: 1–3.

Shen, Xue-Song, Qi Su, Zhuang-Ping Qiu, Jing yi Xu, Yan-xia Xie, Han-fu Liu, and Yi Liu. 2010. "Effects of Artemisinin Derivative on the Growth Metabolism of Tetrahymena Thermophila BF5 Based on Expression of Thermokinetics." *Biological Trace Element Research* 136 (1): 117–125.

Su, Xin-zhuan, and Louis H. Miller. 2015. "The Discovery of Artemisinin and Nobel Prize in Physiology or Medicine." *Science China Life Sciences*: 1175–1179.

Szlezak, Nicole A. 2012. *The Making of Global Health Governance: China and the Global Fund to Fight AIDS, Tuberculosis, and Malaria*. New York: Palgrave Macmillan.

TechnoServe. 2007. "Supporting Artemisia Annua Cultivation in East Africa: Technical and Financial Report. January–December 2006." Tanzania: TechnoServe.

Tibandebage, Paula, Samuel Wangwe, Maureen Mackintosh, and Phares G. M. Mujinja. 2016. "Pharmaceutical Manufacturing Decline in Tanzania: How Possible Is a Turnaround to Growth?" In *Making Medicines in Africa: The Political Economy of Industrializing for Local Health*, edited by Maureen Mackintosh, Geoffrey Banda, Paula Tibandebage, and Watu Wamae, 45–64. International Political Economy. London: Palgrave Macmillan.

Van der Geest, Sjaak. 2011. "The Urgency of Pharmaceutical Anthropology: A Multilevel Perspective." *Curare* 34: 9–15.

Van der Geest, Sjaak, Susan Reynolds Whyte, and Anita Hardon. 1996. "The Anthropology of Pharmaceuticals: A Biographical Approach." *Annual Review of Anthropology* 25: 153–178.

Von Freyhold, M. 2008. "Downstream Issues Regarding the Production and Distribution of ACTs." www.mmv.org/IMG/pdf/6_Downstream_issues_for_ACTsM_von_Freyhold.pdf.

WHO. 2010. *WHO Global Malaria Programme: Good Procurement Practices for Artemisinin-Based Antimalarial Medicines.* Genf: WHO.

WHO. 2016. "Overview of Malaria Treatment." www.who.int/malaria/areas/treatment/overview/en/pdf.

Wright, Codina W. 2002. *Artemisia: Medicinal and Aromatic Plants: Industrial Profiles.* London: Taylor & Francis Groups.

Youyou, Tu. 2011. "The Discovery of Artemisinin (qinghaosu) and Gifts from Chinese Medicine." *Nature Medicine* 17 (19): 1217–1220.

Yu, Hongwen, and Shouming Zhong. 2002. "Artemisia Species in Traditional Chinese Medicine and the Discovery of Artemisinin." In *Artemisia: Medicinal and Aromatic Plants: Industrial Profiles*, edited by C. W. Wright, 149–157. London: Taylor & Francis Groups.

7 Circumventing regulation and professional legitimization

The circulation of Chinese medicine between China and France

Simeng Wang

A large number of people (practitioners, drug sellers, teachers), products (herbal pharmaceuticals, plant and animal raw materials, healing tools), and public and private institutions contribute to the transnational circulation of Chinese medicine. A series of studies about the "worlding" of Chinese medicine (Zhan 2009) were conducted in translocal settings such as the United States (Furth 2011; Pritzker 2012), Tanzania (Hsu 2002) and Norway (Sagli 2009). These authors explored the complexity of the creation and re-creation of so-called "Chinese" knowledge in translocal settings (Zhan 2009). Given the differences in healthcare systems and models of integration in the societies mentioned, a diversity of practices and situational junctures shape doctor and patient behaviors (Scheid 2002). The link between Chinese medicine as it is practiced abroad and the healthcare systems that regulate it in different countries raises the more general question of the nature of the relationship between the circulation and regulation of Chinese medicine in a globalized world. How are the circulations of this medicine structured by governance modalities in translocal networks and contexts? Can we observe the practices that aim to circumvent regulation frameworks during the circulation of this medicine? And what are consequences of this on regulatory systems?

Based on an ongoing empirical study of the transnational circulation of Chinese medicine between China and France,[1] this chapter aims to show, on the one hand, how the circulation of Chinese medicine has been rendered possible between these two countries in the context of the regulation of the last decades, and on the other hand, how this circulation shapes regulatory framework. The central question concerns the interactions between circulations and regulations of this medicine. The goal is to shed light on the diversity of regulatory patterns related to different national contexts, in which practitioners build – each in their own way – their professional legitimacy while circumventing the regulatory framework. By following practitioners' trajectories and legitimation, this chapter proposes a deconstruction of the dichotomous paradigm that frames the relationship between circulation and governance, and analyzes the way these two factors influence each other. The focus will be on the roles played by the actors in the transnational circulation of Chinese medicine: first, the drug sellers involved in the circulation of products, and then, practitioners engaged in the circulation of practices, techniques, and knowledge. This chapter examines the ways by which various actors appropriate the norms imposed by the regulatory system for themselves and adjust their professional practices accordingly. Concurrently, it looks at the consequences of these practices on the regulatory framework and on the changes in governance of Chinese medicine on the global scale. The trajectory approach – following practices and mobility of actors – proves useful in the examination of "local development" in the global diffusion of Asian medicine (Hsu 2012; Pordié and Simon 2013), since individual practices reveal the more general social conditions structuring the development of Asian medicine (Langwick et al. 2012): the stakes of jurisdiction, professional legitimacy/illegitimacy, the marketing of medical products, knowledge transmission, and transnational networks.

The policies of the Chinese government on the promotion of this medicine in the global arena have evolved over the last decades, especially after the Nobel Prize for medicine was awarded to the Chinese scientist Tu Youyou in October 2015. Since the end of 2015, the Chinese government has implemented new public policies to encourage the international development of Chinese medicine. This has offered, through various international and intergovernmental agreements, a renewed context for the circulation of this medicine. In order to be positioned on the already highly regulated market of healthcare in French society, Chinese medicine practitioners and drug sellers are required to respect both international – especially EU (European Union) – rules on the exercise of complementary and alternative medicine (CAM), and French national rules governing the development of Chinese medicine. Based on qualitative materials – interviews, consultation of archives and institutional documentation, and observations – collected since 2014 in three cities (Wuhan, Beijing, and Paris), my work shows that behind all formal and institutionalized forms of promoting Chinese medicine in French society, there are also the illegal practices of ordinary actors, who participate actively in the transnational circulation of this medicine. Those illegal practices, often invisible and neglected, are merit a closer look.

In this chapter, I study on different scales (micro, meso and macro): 1) the diversity of regulators, 2) the variety of practices to contend with the regulatory frameworks, and 3) the process of modifying the regulatory regimes through the practices and trajectories of these actors. The underlying logics of the interdependent relationships between the circulation and regulation of Chinese medicine are based on several elements. First, on the different public and private actors involved in the process of circulation. Second, on the interlinking/interaction of multiple sectors (economic, political, cultural, and scientific) in the various contexts of regulation that influences the agency of stakeholders who experience geographic mobility and develop strategies to circumvent regulations. And finally, on the idea of temporality in analyzing the dynamics of the interdependence between circulation and regulation.

Contextualizing Chinese medicine in France: European and French national regulations

Sinologists agree that there are now five branches of Chinese medicine: acupuncture, Chinese pharmacopeia, Chinese dietetics, Tuina massage, and psycho-energetic practices such as *qi gong* (Cai 2000; Farquhar 1995). In France, as early as 1945, homeopathic physicians were offering one of these – acupuncture – to their patients. As such, the practice of acupuncture is recognized by the central authority, the *"Conseil National de l'Ordre des Médecins,"* and is eligible for reimbursement by medical insurance, making it a "medical act" protected by the law, even though it is based on an illegal practice of medicine (Guilloux 2006; Candelise 2008). Acupuncturists who have followed a different course from the one taken by French acupuncture doctors – even those who have graduated in Chinese medicine or in conventional medicine[2] in their home country – are legally prohibited from practicing in France (Carricaburu and Ménoret 2004). A doctor in France must comply with four conditions: one must hold a medical degree or the equivalent of an MD obtained in a member state of the European Community, have French nationality or be an European Economic Community (EEC) citizen, be registered at the *Conseil National de l'Ordre des Médecins*, and register the degree at the prefecture (according to the health code of the year 2000, Article L. 4161–1). Consequently, a person qualified as a doctor in a country outside the European Union is not allowed to exercise in France without passing a medical degree in France or the authorizing tests to exercise. This application must be completed by three years of work experience in a French hospital. In addition, one must qualify for the state-recognized status of "physician acupuncturist" with two years of study after (or in the course of) graduating in medicine. Doctor acupuncturists thus combine all the most legitimate aspects of the medical and legal norms, and this is why they are able to control the professional jurisdiction.

However, apart from university training in acupuncture alone, some private schools in France offer acupuncture courses alongside classes on pharmacopeia, massage, and dietary and bodily practices (such as *Taiji* and *Qigong*). Dating back to the 1980s, these schools are primarily led by French practitioners who trained in China in the institutes of Chinese medicine that opened to foreigners after the Cultural Revolution in 1976. Gradually, the graduates from these schools began to practice Chinese medicine despite the risk of being convicted for "illegal practice of medicine." Many of them have since created training centers, mostly for the French middle or upper classes, of whom many already have a professional activity. In parallel, there are also schools led by practitioners of Chinese origin. However, most of these practitioners of Chinese medicine in France have not trained at French schools; instead, they obtained practical training in China, some in the specialized centers and the others in the universities. The types of transgression of the legal norms by these different practitioners – in their medical exercise and in their training processes – leads me to identify two main "sub-cultures" (in the sense of Howard Becker [Becker 1985 {1963}]) within the heterogeneous group of Chinese medicine practitioners working outside the conventional framework.

In the late 20th and early 21st centuries, the international trade of pharmaceutical raw materials was facilitated by the developments in biomedical pharmacology (Gaudillière 2002) using so-called "traditional" American and African drugs for health and economic profit purposes. Today, Europe and the United States are still the key actors in setting medical norms (Chorev 2012) at the international level. In addition, international and supranational organizations, such as the World Health Organization (WHO), and the US Food and Drug Administration[3] are involved in the manufacturing and distribution of sanitary norms globally, and also play a significant role in the regulation of Chinese medicine, particularly in terms of Chinese pharmacopeia. Specifically in the case of Europe, in May 2011, the European Union banned all Chinese pharmacopeia based on medicinal herbs that are not registered in the Traditional Herbal Medicinal Products Directive (2004/24/EC).[4] Since this date, the Chinese government has been negotiating with the Committee on Herbal Medicinal Products (HMPC) to promote the commercialization of Chinese pharmacopeia in the Schengen area.[5] However, despite all these regulatory standards, the sale of Chinese pharmacopeia products continues in shops, in less formal sale points, and online. In Paris, these transactions mainly pass through Chinese merchants involved in the import and export of Chinese medical and paramedical products. Some of them work illegally.

This study of the wide variety of circulations and non-regulated practices in Chinese medicine in France aims to show the different regimes of regulation (by market, by scientific training, etc.) and the different ways of legitimizing Chinese medicine in the French context, depending on its content and form (products or services), but also on the various resources possessed by the actors. It raises the question of how collective and/or individual voices in this professional territory are diffused by Chinese medicine actors (practitioners, teachers, and traders) of different nationalities, social origin, ethnic origin, and professional backgrounds. This allows us to study the plurality of circulation itineraries and legitimation processes that are mobilized by actors in response to the regulatory frameworks concerning Chinese medicine in the French national context. This leads us to the question of characterizing the "governmentality" of the body (Foucault 2004) with a multi-scale perspective, and understanding the production of actions and public policies by multiple actors with plural profiles (Lequesne 2001; Brenner 2004).

Cross-border sales of Chinese healthcare products: category changes

In the field of health, transnational exchanges have been bolstered by the development of a "circular economy" (Dubuisson-Quellier 2014) through the multiple circulations of medicines (Pordié 2011, Dudouet 2002, Hauray 2006; Gaudillière 2010; Baxerres 2011). The transnational circulations of drugs are increasingly framed by "good practice" guidelines and by binding standards (Pordié and Gaudillière 2014a). This necessary standardization elicits a more "efficient" circulation of knowledge between different actors (Star and Griesemer 1989). In the past decade, increasing attention has been paid in the social sciences to the regulation of drug policies and markets in a global context. This interest is mainly due to the fact that the regulation procedures and the imposition of national and international standards appear as central sites of power (Petryna et al. 2006; Kumar and Dua 2016). The regulatory framework crucially determines the sale, use and distribution of traditional medicine products. The WHO and the European Union define the basic criteria for evaluating the quality, safety, and efficacy of herbal medicines with the goal of assisting national regulatory authorities, scientific organizations, and manufacturers in assessing documentation, submissions, and dossiers concerning such products (Ajazuddin and Shailendra 2012). This makes it possible, in the flux of transnational circulations, for the complex herbal formulas of Asian medicine to meet the requirements of the European regime of traditional herbal medicinal products (Schwabl and Vennos 2015). Moreover, different countries have adopted various approaches to licensing, dispensing, manufacturing, and trading when regulating traditional medicines.

In Europe specifically, the Traditional Herbal Medicinal Products Directive (THMPD) of April 2011 stipulated that all herbal products not considered as food (such as spices) must have a THMPD authorization to be on the market. The directive also requires that all preparations are subject to the same procedure as chemical drugs. Therapeutic products from China – and from India, Japan, or any other origin outside the European Union – that have not been studied in a clinical trial within the European Union are excluded from the market. Thus, the THMPD authorizes only three categories of herbal medicinal products on the EU market: 1) products that are licensed and have been tested clinically, 2) products that are described as "traditional" by the THMPD and have been trialed, and 3) products that claim to have therapeutic virtues but are not used for medical purposes.

In response to this regulation framework (the ban on Chinese pharmacopeia based on medicinal herbs not registered under THMPD), the Chinese government has been negotiating with the HMPC to support the commercialization of the Chinese pharmacopeia in the Schengen area. Following the successful registration of the first Chinese medicine product named *"di'ao xinxuekang"*(地奥心血康) as a traditional medicine via the Medicines Evaluation Board of the Netherlands, several other traditional Chinese medicine herbal products are undergoing the registration process in various European member states (Wu et al. 2015). Meanwhile, the Chinese pharmacopeia producers seeking to position themselves on the European market have adopted a new manufacturing strategy: they class products that are patented and sold in pharmacies as drugs in China under a different category in the European Union, as "functional food." Thus, we can observe two processes: one whereby the regulation framework is altered by politics (negotiations between the Chinese government and international organizations), and another whereby it is circumvented by economic and pharmaceutic actors who are driven by the market (Pordié and Gaudillière 2014b; Elaheebocus and Mahomoodally 2017).

This new "functional food" category originated in Japan in 1995, then developed in the United States before reaching Western Europe. This neologism is formed by combining the two concepts – food and drug. Sometimes referred to as fortified foods, this category creates direct competition between the food industry and the pharmaceutical industry (Bardou-Boisnier and Caillaud 2015). These are standard foods artificially enriched with nutritional compounds either during the production process (eggs or meat with omega-3) or during the transformation process (dairy products enriched with vitamin D, margarine or butter containing phytosterols). The phrasing implies that they are relatively safe and that everyone can theoretically consume them with no risk. Through the consumption of these food products, we can observe the process of medicalization alongside an increasing interest in dietetics.

The case of Chinese herbal tea (*liang cha* 凉茶, literally translated as "cold tea") is also relevant. Dating back to the 3rd–5th century Eastern Jin dynasty, the "Handbook of Prescriptions for Emergencies" ("肘后备急方," edited by Ge Hong) recorded Chinese herbal recipes to reduce excessive heat in the body that were often used in the south of China's Mountains (Ling Nan area). These herbs are said to be capable of reducing excessive heat inside the body and are used for diseases primarily characterized by inflammation or infection. The concept of heat in Chinese medicine is believed to be the cause of many "hot symptoms." Herbs that can reduce "fire heat" are used to treat acute inflammatory and infectious diseases, because such diseases can be said to attack as swiftly as a fire burning down a building. They were originally called "cold drugs,"[6] but Cantonese people disliked the presence of the word of "drug" in the appellation and therefore substituted the word "tea." After the 19th century, "cold tea" was sold less and less in the streets. Since the end of the 1980s, many Chinese pharmaceutical companies have used the traditional formula of "cold tea" to develop a drug, named "isatis root granule" (*banlangen chongji* 板蓝根冲剂). This pharmaceutical, over-the-counter drug, is mainly sold in pharmacies in contemporary China.

This product is prohibited under the THMPD and so cannot circulate as a "drug" within the Schengen area. However, it can be found on the drinks/tea shelf in most Chinese supermarkets in the Paris region and also at points of sale of Chinese herb plants. The term "drug" does not appear on the packaging; instead, the word used to describe the nature of the product is "beverage"; whereas on the packaging of Isatis root granule sold in China, there is a bar code alongside the phrase "electronic regulation code of Chinese drugs" (*zhongguo yaopin dianzi jianguan ma* 中国药品电子监管码). This change of terms used on the packaging reflects the strategy developed by pharmaceutical firms for circumventing international regulations by changing the category of the commercialized products.

Following is an extract of an interview with Mr Zhou, owner of a Chinese herbal plant point of sale. It sheds light on the possibilities for the circulation of Chinese pharmacopeia from China to Europe, and the category change of these products from "drug" to "drink (food)."

> In our profession, we have to learn to "deal with it." The European Union says it is not a drug so we have to find a way to qualify it differently. "Care products" appear as a great opportunity for us. In this box, we can put so many Chinese medicine products, calling them functional foods, like plants, spices, especially some of the raw materials.

As Mr Zhou said, the circulation of certain Chinese pharmacopeia is made possible by changing the product category. For those products that cannot be categorized as "functional foods," another way of circumventing regulations is through illegal trading. In this regard, the trafficking of seahorses, often by smuggling them in suitcases, is revealing of the underground economy of Chinese pharmacopeia in France. In a different context, Lee Mendoza (2009) evidenced the link between traditional and alternative medicines practices and the informal health economy in the Philippines.

To enable the transnational circulation of Chinese medicine products, the traders I interviewed have developed their own personal commercial strategies to bypass regulation. But when it comes to providing a service, as is the case with Chinese medicine practitioners during their medical appointments/consultations, the modalities of their circumvention of regulation frameworks are more related to legitimacy building and to professionalization processes.

Numerous studies have been focused on professional legitimation among traditional medicine practitioners: traditional medicine practitioners in Senegal (Fassin and Fassin 1988), CAM practitioners in Canada (Kelner et al. 2004; Ijaz et al. 2015), acupuncturists in Britain (Givati and Hatton 2015), and CAM practitioners in Australia (Wiese and Oster 2010). Previous works have made it clear that beyond alliances with actors in the market and in other professions, the importance of an alliance with one part of the state provides a certain stability and autonomy to the profession (Abbott 2003). In the next sections, two professional situations will be distinguished. The first relates to illegal practitioners already working in the profession, and the second refers to the training of future practitioners before they enter the Chinese medicine profession. These two situations show the different processes of legitimation. Some illegal practitioners already practicing gain legitimacy by campaigning and advocacy, and others circumvent the French regulations through their own geographic mobility. As for the training of future practitioners, Chinese medicine teachers are increasingly seeking to influence the control of professional territory.

Illegal practitioners: legitimation by campaigning or circumvention by geographic mobility

Although practitioners from non-medical professions are all categorized in France under the "illegal practice of medicine" category, they nonetheless experience very different degrees of job insecurity. There is a huge gap between a French senior official holding a PhD in social sciences who has trained in Chinese medicine in China and is practicing in a provincial town, and an undocumented Chinese migrant providing acupuncture treatments to his compatriots in Paris. With respect to the issue of illegality, both in terms of interactions between the French medical professions and the state – through the Ministry of Justice, but also the Ministry of Health (Ministry of Solidarity and Health since 2017) – and as concerns the rights of Chinese migrant practitioners in the host society, the professional practice of Chinese medicine in France is a very segmented space.

Over the last ten years, some practitioners in Chinese medicine – almost exclusively Europeans of non-Chinese origin (French, Belgian, and Swiss) – have joined forces to try to establish supervisory and homogenization criteria for training and professional practices (Parent 2014). The members of the major federations, unions, and professional associations representing Chinese medicine practitioners pursue their professional careers in France, even though they frequently spend time in China. Lawyers for these professional organizations use European law in order to construct a status and a professional territory. For example, they are inspired by recent regulations adopted in other European countries, such as Portugal, which passed a law in 2013 regulating the practice of Chinese medicine by non-physician practitioners. In France, the Strategic Analysis Centre attached to the Prime Minister's office published a briefing note summarizing the different European statuses of practitioners and suggested a "label for therapists of unconventional practices" be established (Reynaudi 2012, 1). In this sense, we can speak of a "visibilization" of illegal practitioners and their professional practices in the legal sphere over the past decade. And more recently, this legitimation has also been discussed in the media.

However, practitioners of Chinese origin are poorly represented in these organizations. They seldom assemble in associations, and even if they do, they do not voice professional demands. How can this "invisibility" of practitioners of Chinese origin in French public debates be explained? One of the first answers is linked to the double illegal status of some of them: they are both undocumented residents and illicit practitioners. As Mr Chen says, they prefer to "shut up" and "keep quiet." Even the legal migrants prefer to keep their practice discreet. Therefore, the relationship of these practitioners with the state can sometimes be experienced as a constraint limited to administrative formalities regarding the obtention of the medical diploma, or the illegality of their personal status. However, all Chinese-speaking practitioners of Chinese origin we met use virtual communities on a daily basis. On these internet and social networks, they discuss "good" and "bad" professional practices of Chinese medicine in a French context: how to find clients, the techniques to "retain" them, how to avoid being "spotted" and "denounced." These online forums allow them, on the one hand, to get to know each other and to maintain a professional network, for example by posting announcements of recent openings in medical practices; and on the other hand, to stay up to date on the latest news not only scientifically – including news and breakthroughs in Chinese medicine published in China – but also political guidelines and regulation policies in France and more widely in Europe.

Regardless of the national origin of the practitioner, it is through professional networks that these illegal practitioners learn to circumvent the regulation of Chinese medicine practice in France. Some of them rely on the European clauses and legally practice Chinese medicine in neighboring countries where the profession is regulated, such as Switzerland and Spain; others simply go offshore. Indeed, farther than 15 kilometers offshore is no longer considered France. The mobility of practitioners beyond the French territory has become one of the ways to avoid controls and judiciary attacks on illegal practice in Chinese medicine. To do this, a series of practices allowing the geographical mobility of practitioners and also of their clients are being implemented. This quote from Madam Goffroy, an acupuncturist-naturopath living in Brest, is revealing:

One of the benefits of being in Brittany is the sea! Some of us [illegal prac-titioners] choose to exercise off the coast. But that needs some planning: one must have a boat, a permit and be able to motivate clients to get on a boat. It is an entire organization and represents an extra cost!

Apart from this solution of sailing out to sea, other practitioners choose to cross borders and set up in neighboring countries. The experiences shared by Mr Cheng testify to the margins of professional jurisdiction as a result of practitioner mobility:

To settle in Switzerland, I first had to present a certificate of my training hours. I showed my degree obtained in a Chinese Medicine University [in China], and my certificates for a few internships completed in France. Then I contacted the Swiss organizations involved in validating foreign therapists. Some are run by doctors, others by non-doctors. Considering my experi-ences, dealing with organizations run by doctors is harder. Finally, it was an organization of non-doctors that validated my status.

Furthermore, to sustain their activity, the non-doctor practitioners interviewed develop the capacity to create cross-professional alliances (Abbott 2003) with other actors in the field of Chinese medicine, such as merchants and sellers. This is the case for Ms Li, a Chinese medicine practitioner of Chinese origin practic-ing in a working-class district of Paris. She collaborates with an herbalist in her neighborhood:

The herbalist's store is at the end of the street. They are our partners! The couple settled before me, but we come from neighboring villages [they all come from Zhejiang province, China]! So very quickly we became friends. We help each other and it is a win-win relationship between us. They talk about my clinic to their clients and I like to advise my patients about buy-ing herbs products at their place! . . . In this immigrant neighborhood, there are more and more Chinese people. They are our main clients. Neither the herbalist nor I expect very much to earn money with the French people! [Laughter]

When practitioners do not have legal status, building customer loyalty relies even more on these cross-professional alliances, and word-of-mouth recommenda-tions prove to be crucial. And in this case of a win-win relationship between two Chinese migrants, a practitioner and an herbalist, internal regulation of the pro-fessional group operates through the mobilization of personal networks, where the ethnic criterion come into play. This finding falls in line with Lieber's work (2012) on Chinese medicine practitioners – of Swiss and Chinese origins – in Switzerland who mobilize two very different forms of cultural legitimacy, as well as with Chiu's (2006) study of Chinese immigrant practitioners of traditional Chi-nese medicine residing in British Columbia, Canada.

Another way for illegal practitioners to circumvent the French regulation frameworks takes place before they enter into the profession, during their training. As the next section will demonstrate, some Chinese medicine teachers seek to influence the professional territory controlled by the French state by advocating the establishment of personal, professional, and institutional links with European and Asian countries, especially China.

Circumventing French regulation through international norms and training collaborations

A considerable number of the teachers of Chinese medicine in France are Chinese, or of Chinese origin. Their international circulation – through professional mobility and/or a migratory trajectory – contributes to promoting the recognition of Chinese medicine by the French state. Similarly to the practitioners of non-Chinese origin, despite their illegal status in France, they step onto the international stage in order to legitimate themselves at a national level.

In order to gain professional legitimacy, the professional bodies initially sought to forge state alliances within the Ministry of Health. They took several approaches, including explaining the safety of the practices, ensuring the quality of the training, and the control of practitioners through the implementation of professional standards. On this basis, organizations are now lobbying for a regulation that would prevent judicial proceedings for illegal practice of medicine. However, these professional bodies have failed in their endeavor despite a decade of discussions with Ministry of Health advisors, who often are physicians. Their quest for allies within the state was rendered even more difficult by the prior success of medical acupuncturists in institutionalizing their practice within the medical professions.

Thereafter, since it proved impossible to find allies within the state, the professional organizations have bypassed the administrative machinery (the Ministry of Health in particular) and sought to mobilize public opinion instead. Two types of approaches can be identified. The first is to ask practitioners to contact their local representative/MP to expose the injustice of their situation, in order to bring the matter to the National Assembly. This circumvention of healthcare authorities by seeking political support is based on a set of new possibilities offered by European standards, which complement the professional legislations in France today. They include a proposal for a "therapist label" based on the model of the *Heilpraktiker* in Germany and the establishment of an ISO 9001 (international standards organization) standard for the creation of a new healthcare profession.

The second approach mobilizes a network of practitioners to develop political contacts who can make the voice of unrecognized practitioners heard at both French and European levels. These latter practitioners are often connected with academies and schools of Chinese universities, and rely on personal connections to establish cooperation and certifications which grant them extra credit on the professional market – even if those diplomas are not officially recognized under current French legislation. This is, for example, the case for the director of a school of practitioners who made his first trip to China at the end of the 1980s. He has since returned there every year to visit several Chinese hospitals for between three weeks and a month. His school now benefits from this experience that allows him to deliver certificates to his students, stamped by the World Federation of Acupuncture Society (WFAS) and attached to the World Health Organization. The mobilization of this international order to gain legitimization in the eyes of French state allows us to extend the Andrew Abbott's analysis of the steps taken by professionals, within political ecology, in the demarcation and the maintenance of a professional territory (Abbott 2005). We now turn to an analysis of the case of Mr Cai.

Mr Cai is a doctor in Chinese medicine; trained and licensed in China, he acts as a research engineer in the integrated center of Chinese medicine created by the administration of Paris Hospitals. This center is one of a kind and was founded as part of the 2010–2014 Paris Hospital strategic plan to coordinate and evaluate the scientific and economic interest of Chinese medicine practices within public hospitals in France. Mr Cai is one of the practitioners of Chinese origin who have been asked to manage official relations with China, including "clinical assessments," the term used by the center. Although this incorporation of Chinese medicine into hospital practices – institutions with strong legitimacy in the French medical space – appears to confirm a strong interest for Chinese medicine, the center is used to support existing practices and "evaluate the effectiveness of Chinese medicine," and its goal is not in fact to expand the professional territory of Chinese medicine practitioners. Furthermore, although the strategic plan 2010–2014 facilitated the development of scientific cooperation between Parisian and Chinese hospitals, the funding for this project is not renewed in the strategic plan 2015–2019. According to Mr Cai, the reasons given by the members of the Center are: the "necessary" cuts in public expenditure and the opposition mounted by a doctors' union against this integration of Chinese practices. For Mr Cai, this power relationship is closely linked to the specific situation of the professional practices of Chinese medicine in France:

> It is possible to distinguish three main groups of Chinese medicine practitioners in France. First, public health service physicians (*médecins conventionnés*) who practice Chinese medicine. They do not consider non-physicians, nor physicians of Chinese medicine trained and licensed in China, as legitimate practitioners. The second group is composed of private physicians, representing the largest share: I think 80% of the practitioners are of this kind. And finally, the experts I would say, who are in my eyes the "real" doctors of Chinese medicine in France, they are Chinese! Born in China, educated and trained in China, they read ancient volumes of Chinese medicine directly in the original language, not in their English or French translations. Since they are highly segmented within the so-called "Chinese medicine practitioners," it is very difficult to unite these three groups.

Including himself as one of the "real" doctors of Chinese medicine – Chinese speaking and trained in Chinese universities – Mr Cai here adds value to his "authentic expertise" (Lambert 2012) and highlights the existence of a powerful medical monopoly which is closely linked to the possession of the French medicine qualification. Indeed, in France, physicians are able to maintain tight relationships with the state regarding the control of professional territory, with a privileged access to the Ministry of Health (Hassenteufel 1997). The dominance of biomedicine in establishing educational standards, regulation and market control of complementary and alternative medicine is also evident in other countries, such as Canada (Shahjahan 2004).

In fact, collaborating with other countries, especially with China, has become a means of legitimizing one's practice vis-à-vis the French state. The interview with the director of the center of Chinese medicine, Mr Durand, highlights his strong desire to create links with China:

> Even if we have not been granted the funding in the new strategic plan 2015–2019, our activities are sharply on the rise. Thanks to networks created with Chinese colleagues in several Chinese cities, there are more and more students and practitioners in France coming to us to be trained. In France, we also act as the primary spokespersons for several ministries: Foreign Affairs, Culture, Healthcare. Because we have been assigned this role as the interface between France and China in the field of Chinese medicine. Our experiences of working with senior Chinese officials demonstrate our expertise! The Chinese government invests more and more in the promotion of Chinese medicine internationally. For us, it's an opportunity!

Indeed, the attention paid by the Chinese government to the international development of Chinese medicine increased after the Nobel Prize was awarded in October 2015 to Tu Youyou, a Chinese pharmacologist who contributed to research treating malaria with artemisinin. Even before the award, according to Ma Jianzhong (China Acadamy of Chinese Medical Sciences n.d.), the deputy director of the national administration of Chinese medicine and Chinese pharmacopeia, Chinese medicine was present in 183 countries and regions of the world and Confucius Institutes specifically dedicated to the dissemination of Chinese medicine were built in four countries. There were also 83 "cooperation in Chinese medicine programs," signed between China and other governments or between China and transnational organizations. Additionally, acupuncture, one of the most common branches of Chinese medicine in allopathic clinics, was registered as an Intangible Cultural Heritage of Humanity by the United Nations Educational, Scientific and Cultural Organization (UNESCO) on 16 November 2010. According to WHO statistics (Organisation mondiale de la Santé 2013, p. 22), acupuncture is recognized by health insurance schemes in 18 countries and regions.

Despite this significant presence on the international scene, the Chinese government became more conspicuously involved in internationalizing Chinese medicine after Tu Youyou's prize was awarded. They introduced a range of measures to attract more foreigners seeking to be trained in Chinese medicine in China, ranging from initial training with graduation to continuous training courses and clinical internships. Both illegal and legal French practitioners seized these opportunities to create exchange and cooperation programs with China. The words of Mrs Lacré, director of a training school located in the Department of Rhône-Alpes, are revealing:

> In the region, what distinguishes our school from others is the links that we have forged with China. Thanks to Professor Luo, a practitioner of Chinese medicine in Kunming, we have started an Exchange program with the Faculty of Chinese medicine of Kunming. In our training courses for students, a trip to Kunming and 50 hours of in-class sessions at this school are included. Our training is authentic in the sense that there are Chinese teachers and that we learn Chinese medicine in the country where it comes from. Our students are elated. We are now engaging in another program, supported by the Chinese authorities and set as a priority by the local government. This program is more focused on medicinal plants. The idea is to present the process of treating the cultures, harvest and plants to the students.

As Ms Lacré says, the Chinese state and administrations at different scales are now implementing institutional and public policies aiming to expand the international influence of Chinese medicine, through agreements and the training of future practitioners. In this context, Chinese medicine (like other Asian medicines) is not only promoted as an international commodity, but it is also associated with its philosophical visions of the world and the discovery of cultures and unique civilizations (Pordié 2011). These labels of "exotic" and "natural" attributed to Chinese medicine match the expectations of foreign students, and offer niche Chinese medicine training opportunities abroad: instead of publicizing their courses as a curriculum of teaching practices and medical techniques, some training schools highlight the spiritual, philosophical, and cultural dimensions of this medicine. Such international collaborations thus meet the expectations of the institutions and foreign practitioners as well as the Chinese government incentives.

Mr Hu, the director of a training center, himself practices Chinese medicine illegally in the south of France. His account illustrates a mechanism of professional legitimation in the transnational space; in other words, the ways of making use of the resources present in the different areas of regulation through the circulation of models and ideas:

> In our center, we do not only deliver expertise in Chinese medicine, but also discuss the cultural and philosophical thinking related to it. In fact, it is on these last points that we base the promotion of our training. We are inspired by the model of the Confucius institutes. The idea is to create a place to experiment with Chinese medicine through cultural activities – conferences, visits, video projections, and training – related to dietetics, aesthetics, massage, acupuncture, and pharmacopoeia. If I take the term used by Chinese policy makers, it is a form of Chinese medicine "soft power" that we strive to spread abroad! With this idea as the guideline for the development of our center, we drew the attention of our Chinese counterparts willing to work

with us. In France, these international networks and resources are of course very rewarding!

The term "soft power"[7] is widely used by the Chinese authorities in association with the diffusion of Chinese influence abroad. The "internationalization" (term used by the Chinese state) of Chinese medicine relies on "strengthening the two powers": the "soft power diffusing cultures of Chinese medicine abroad" and the "hard power," meaning innovative research on technology and techniques of Chinese medicine (Xinhua Health 2017).[8] In order to extend the influence of Chinese medicine, the media and training programs abroad are encouraged to target various interested parties: economic actors interested in investing abroad, practitioners overseas, and foreign devotees of Chinese medicine and cultures. Mr Hu was able to create a niche for the development of his center in France by seizing the opportunities offered by the political incentives from the Chinese government. The establishment of close ties with the Chinese government brought him closer to practitioners and institutions located in China. Thereafter, the networks built between Mr Hu and his Chinese partners made his center more authentic and legitimate in the eyes of his French students and other practitioners in the French professional territory. Here is an example of how scientific, social, and administrative resources gathered on the international arena can be mobilized to legitimize practices considered "illegal" at the national level (in the case of France).

Conclusion

By studying the circulation of Chinese medicine between China and France, I have highlighted some key elements in the relationship between transnational circulation and the governance of Asian medicine. More precisely, this chapter has focused on the practices and strategies adopted by product sellers and illegal practitioners to circumvent regulatory rules and gain professional legitimacy. Given the diversity of content of Chinese medicine and its multiple forms (acupuncture, pharmacopeia, dietetics, massage, and psycho-energetic practices), it was important in this analysis to take into account both circulations of products and that of techniques/practices. The latter is divided into three sections, respectively: the sale of Chinese medicine products, the professional legitimation of illegal practitioners, and the training conditions of future Chinese medicine practitioners.

In each section, the findings illustrate how circulation and regulation affect one another. In the sale of Chinese medicine healthcare products, traders specialize in informal import-export and position themselves in the gray areas of regulation. They change the status and category of the products commercialized to ensure the circulation of these products categorized as "illegal" by European and French regulation frameworks.

In the case of illegal practitioners, some of them – almost entirely practitioners of French origin – try to legitimate their practices by campaigning through associations that stand up to French physicians and decision-makers. While others choose geographic mobility as a way of circumventing French regulation. They leave the French territory and practice somewhere else (at sea or in neighboring countries) where the regulatory pattern is different. Thus, the way practitioners build their own practices is heavily influenced by the regulatory frameworks.

Concerning the training conditions of future practitioners, teachers of Chinese medicine seek to circumvent French regulation using international norms and transnational collaborations. They aim to influence the professional territory controlled by the French state, either through applying the international norms, or through establishing personal, professional, and institutional links with countries in Europe and Asia, especially China.

As a whole, my study demonstrates the different ways various actors circumvent regulations in the course of the transnational circulation of Chinese medicine. There exist tight and diversified links between the ways the actors circumvent governance and how they legitimize themselves. Each actor individualizes their legitimating strategies according to their professional activity and the resources they possess. The various different regulation frameworks create highly differentiated itineraries of Chinese medicine circulation. Further investigation into this subject would benefit from using the global trajectory approach. Transnational circulation does not only function through formal regulation and official rules. It also passes through unofficial regulation frameworks and through individuals' trajectories and mobility, which deserve serious consideration.

Notes

1 This research has been conducted in the framework of the GLOBHEALTH project, funded by the European Research Council.
2 I use here the term "conventional medicine" as defined by the European Union, and in the negative, I define all medicines and therapies not covered by the dominant European norms as "unconventional medicine."
3 This organization, among others, is mandated to authorize the marketing of drugs in the United States.
4 www.who.int/bulletin/volumes/90/8/12-020812/zh/http://www.who.int/bulletin/volumes/90/8/12-020812/zh/
5 The Schengen area includes the territories of the 26 European States – 22 Member States of the European Union, and 4 associated States, members of European Free Trade Association – which implemented the Schengen Agreement and the Schengen Convention signed in Schengen (Luxembourg) in 1985 and 1990. The Schengen area functions as a single area for international travel and border controls, where internal borders are crossed freely, without passports and without controls.
6 See the "岭南卫生方" of the Yuan dynasty.
7 Here I take the definition of "soft power" as stated by Joseph Nye: "soft power is the ability to affect others to get the outcomes one wants through attraction rather than coercion or payment"; "a country's soft power rests on its resources of culture, values and policies" (2008, 94).
8 Speech by Wei Zichuan, Executive Vice President of the Xinhua website and government spokesperson. See http://www.tcmhsn.com/list/24/246.htm.

References

Abbott, Andrew. 2003. "Écologies liées: à propos du système des professions." In *Les professions et leurs sociologies*, edited by Pierre-Michel Menger, 29–50. Paris: Éditions de la Maison des sciences de l'homme.

Abbott, Andrew. 2005. "Linked Ecologies: States and Universities as Environments for Professions." *Sociological Theory* 23 (3): 245–274.

Ajazuddin, Ajaz, and Saraf Shailendra. 2012. "Legal Regulations of Complementary and Alternative Medicines in Different Countries." *Pharmacognosy Reviews* 6 (12): 154–160.

Bardou-Boisnier, Sylvie, and Kevin Caillaud. 2015. "Les dispositifs informationnels sur les compléments alimentaires: une affaire de sante publique." *Questions de communication* 1 (27): 79–104.

Baxerres, Carine. 2011. "Pourquoi un marché informel du médicament dans les pays francophones d'Afrique?" *Politique africaine* (123): 117–136.

Becker, Howard S. 1985 [1963]. *Outsiders, Études de sociologie de la déviance*. Paris: Armand Colin.

Brenner, Neil. 2004. *New State Spaces: Urban Governance and the Rescaling of Statehood*. Oxford: Oxford University Press.

Cai, Jingfeng (蔡景峰). 2000. *General History of Chinese Medicine* (中国医学通史). Beijing: People's Medical Publishing House (人民卫生出版社).

Candelise, Lucia. 2008. *La médecine chinoise dans la pratique médicale en France et en Italie, de 1930 à nos jours: représentations, réception, tentatives d'intégration*, PhD Diss. EHESS, Paris.

Carricaburu, Danièle, and Marie Ménoret. 2004. *Sociologie de la santé: institutions, professions et maladies*. Paris: Armand Colin.

China Acadamy of Chinese Medical Sciences. n.d. http://www.catcm.ac.cn/zykxyydd/dtyw/201511/17239c0c644f4e2b821dc82020d5a4b6.shtml

Chiu, Lyren. 2006. "Practising Traditional Chinese Medicine in a Canadian Context: The Roles of Immigration, Legislation and Integration." *Journal of International Migration and Integration* 7 (1): 95–115.

Chorev, Nitsan. 2012. *The World Health Organization between North and South*. Ithaca: Cornell University Press.

Dubuisson-Quellier, Sophie. 2014. "Les engagements et les attentes des consommateurs au regard des nouveaux modes de consommation: des opportunités pour l'économie circulaire." *Annales des Mines* 76: 28–32.

Dudouet, François-Xavier. 2002. "L'industrie pharmaceutique et les drogues." *Studia Diplomatica* 55 (5–6): 145–170.

Elaheebocus, Naailah, and Fawzi M. Mahomoodally. 2017. "Ayurvedic Medicine in Mauritius: Profile of Ayurvedic Outlet, Use, Sale, Distribution, Regulation and Importation." *Journal of Ethnopharmacology* 197: 195–210.

Farquhar, Judith B. 1995. "Rewriting Chinese Medicine in Post-Mao China." In *Knowledge and the Scholarly Medical Traditions*, edited by Donald George Bates, 251–276. Cambridge: Cambridge University Press.

Fassin, Didier, and Eric Fassin. 1988. "Traditional Medicine and the Stakes of Legitimation in Senegal." *Social Science & Medicine* 27 (4): 353–357.

Foucault, Michel. 2004. *Naissance de la biopolitique: cours au Collège de France, 1978–1979*. Paris: Seuil.

Furth, Charlotte. 2011. "Becoming Alternative? Modern Transformations of Chinese Medicine in China and in the United States." *Canadian Bulletin of Medical History* 28 (1): 5–41.

Gaudillière, Jean-Paul. 2002. *Inventer la biomédecine. La France, l'Amérique et la production des savoirs du vivant (1945–1965)*. Paris: La Découverte.

Gaudillière, Jean-Paul. 2010. "Une marchandise scientifique? Savoirs, industrie et régulation du médicament dans l'Allemagne des années trente." *Annales. Histoire, sciences sociales* 65 (1): 89–120.

Givati, Assaf, and Kieron Hatton. 2015. "Traditional Acupuncturists and Higher Education in Britain: The Dual, Paradoxical Impact of Biomedical Alignment on the Holistic View." *Social Science & Medicine* 131: 173–180.

Guilloux, Ronald. 2006. *De l'exotique au politique: la réception de l'acupuncture Extrême-Orientale dans les systèmes de santé français (XVII–XXe siècles)*, PhD Diss. Université Lumière Lyon 2.

Hassenteufel, Patrick. 1997. *Les Médecins face à l'État: une comparaison européenne.* Paris: Presses de Sciences Po.

Hauray, Boris. 2006. *L'Europe du médicament. Politique: Expertise: Intérêts privés*. Paris: Presses de Sciences Po.

Hsu, Elisabeth. 2002. "'The Medicine from China Has Rapid Effects': Patients of Traditional Chinese Medicine in Tanzania." *Anthropology and Medicine* 9 (3): 291–314.

Hsu, Elisabeth. 2012. "Mobility and Connectedness: Chinese Medical Doctors in Kenya." In *Medicine, Mobility and Power in Global Africa: Transnational Health and Healing*, edited by Hansjorg Dilger, Abdoulaye Kane, and Stacey Langwick, 295–315. Bloomington: Indiana University Press.

Ijaz, Nadine, Heather Boon, Sandy Welsh, and Allison Meads. 2015. "Supportive but 'Worried': Perceptions of Naturopaths, Homeopaths and Chinese Medicine Practitioners through a Regulatory Transition in Ontario, Canada." *BMC Complementary and Alternative Medicine* 15 (1): 312.

Kelner, Merrijoy, Beverly Wellman, Heather Boon, and Sandy Welsh. 2004. "The Role of the State in the Social Inclusion of Complementary and Alternative Medical Occupations." *Complementary Therapies in Medicine* 12 (2): 79–89.

Kumar, Nandini K., and Pradeep Kumar Dua. 2016. "Status of Regulation on Traditional Medicine Formulations and Natural Products: Whither Is India?" *Current Science* 111 (2): 293–301.

Lambert, Helen. 2012. "Medical Pluralism and Medical Marginality: Bone Doctors and the Selective Legitimation of Therapeutic Expertise in India." *Social Science & Medicine* 74 (7): 1029–1036.

Langwick, Stacey, Hansjorg Dilger, and Abdoulaye Kane. 2012. "Introduction: Transnational Medicine, Mobile Expert." In *Medicine, Mobility and Power in Global Africa: Transnational Health and Healing*, edited by Hansjorg Dilger, Abdoulaye Kane, and Stacey Langwick, 1–30. Bloomington: Indiana University Press.

Lee Mendoza, Roger. 2009. "Is It Really Medicine? The Traditional and Alternative Medicine Act and Informal Health Economy in the Philippines." *Asia Pacific Journal of Public Health* 21 (3): 333–345.

Lequesne, Christian. 2001. *L'Europe bleue. À quoi sert une politique communautaire de la pêche?* Paris: Presses de Science Po.

Lieber, Marylène. 2012. "Practitioners of Traditional Chinese Medicine in Switzerland: Competing Justifications for Cultural Legitimacy." *Ethnic and Racial Studies* 35 (4): 757–775.

Nye, Joseph S. 2008. "Public Diplomacy and Soft Power." *The Annals of the American Academy of Political and Social Science* 616: 94–109.

Organisation mondiale de la Santé. 2013. https://apps.who.int/iris/bitstream/handle/10665/95009/9789242506099_fre.pdf?sequence=1

Parent, Fanny. 2014. "'Seuls les médecins se piquent d'acupuncture?' Le rôle des associations professionnelles de praticiens dans la régulation de pratiques professionnelles de médecine chinoise en France." *Terrains & travaux* 2 (25): 21–38.

Petryna, Adriana, Arthur Kleinman, and Andrew Lakoff, eds. 2006. *Global Pharmaceuticals: Ethics, Markets, Practices*. Durham: Duke University Press.

Pordié, Laurent. 2011. "Savoirs thérapeutiques asiatiques et globalisation." *Revue d'anthropologie des connaissances* 5 (1): 3–12.

Pordié, Laurent, and Jean-Paul Gaudillière. 2014a. "The Reformulation Regime in Drug Discovery: Revisiting Polyherbals and Property Rights in the Ayurvedic Industry." *East Asian Science, Technology and Society* 8 (1): 57–79.

Pordié, Laurent, and Jean-Paul Gaudillière. 2014b. "Industrial Ayurveda. Drug Discovery, Reformulation and the Market." *Asian Medicine* 9 (1–2): 1–11.

Pordié, Laurent, and Emmanuelle Simon, eds. 2013. *Les nouveaux guérisseurs: biographies de thérapeutes au temps de la globalisation.* Paris: EHESS.

Pritzker, Sonia. 2012. "Living Translation in U.S. Chinese Medicine." *Language in Society* 41 (3): 343–364.

Reynaudi, Mathilde. 2012. *Quelle réponse des pouvoirs publics à l'engouement pour les médecines non conventionnelles?* Note d'analyse 290. http://archives.strategie.gouv.fr/cas/system/files/2012-10-02-_medecinesnonconvetionnelles-na290_0.pdf.

Sagli, Gry. 2009. "Learning and Experiencing Chinese Qigong in Norway." *East Asian Science, Technology and Society* 2 (4): 545–566.

Scheid, Volker. 2002. *Chinese Medicine in Contemporary China: Plurality and Synthesis, Science and Cultural Theory Series.* Durham and London: Duke University Press.

Schwabl, Herbert, and Cecile Vennos. 2015. "From Medical Tradition to Traditional Medicine: A Tibetan Formula in the European Framework." *Journal of Ethnopharmacology* 167: 108–114.

Shahjahan, Riyad. 2004. "Standards of Education, Regulation, and Market Control: Perspectives on Complementary and Alternative Medicine in Ontario, Canada." *The Journal of Alternative & Complementary Medicine* 10 (2): 409–412.

Star, Susan Leigh, and James R. Griesemer. 1989. "Institutional Ecology, 'Translations', and Boundary Objects: Amateurs and Professionals in Berkeley's Museum of Vertebrate Zoology, 1907–1939." *Social Studies of Science* 19 (3): 387–420.

Wiese, Marlene, and Candice Oster. 2010. "'Becoming Accepted': The Complementary and Alternative Medicine Practitioners' Response to the Uptake and Practice of Traditional Medicine Therapies by the Mainstream Health Sector." *Health* 14 (4): 415–433.

Wu, Wan-Ying, Wen-Zhi Yang, Jinjun Hou, and De-an Guo. 2015. *Current Status and Future Perspective in the Globalization of Traditional Chinese Medicines.* www.wjtcm.org/ch/reader/create_pdf.aspx?file_no=20140027&year_id=2015&quarter_id=1&falg=1.

Xinhua Health. 2017. Wei Zichuan: Paying Attention to the International Spread of Chinese Medicine Culture. Traditional Chinese Medicine Health Service Network web site: http://www.tcmhsn.com/list/24/246.htm.

Zhan, Mei. 2009. *Other-Worldly: Making Chinese Medicine through Transnational Frames.* Durham: Duke University Press.

Afterword

Governance, circulation, and pharmaceutical objects

Laurent Pordié

It is now time to wrap things up. We learn from this book that circulation and governance relate to a two-fold process. First, circulation is primarily an industry-led phenomenon, firmly anchored in national and global market construction. The globalization of Asian medicine is not the privilege of national and supranational public initiatives, as mainstream components of the global health regime might be (Packard 2016). The World Health Organization, for instance, has shifted its agenda on "traditional medicine" from a policy of integration into primary healthcare – launched during the 1978 Alma-Ata Conference – to a strategy supporting the globalization of traditional medicine through regulation, therapeutic evaluation and safety, property rights and education of professionals' and consumers. While economic reforms and specific government incentives do exist and boost the global development of traditional medicine, as in China or among exile Tibetans (Scheid 2002; Saxer 2013; Kloos 2019), the global turn of Asian medicine is largely private and industrial. The second observation bears on the governance of Asian medicine, which appears to be dominated by public institutions and policies, notwithstanding forms of governance introduced through public-private partnerships (Buse and Walt 2002). Implicit in this opposition is the reciprocal interaction between circulation and governance, their areas of collusion and their cross-fertilization. Not only does the circulation of objects and people inform and inflect the making of governance, but also, modes of governance frame the ways objects and people circulate. All chapters in this book underscore one of these phenomena – or both. Eunjeong Ma shows in Chapter 2 how the Korean government makes use of biomedical science to set up guidelines for public policy and regulatory practices, which in turn set the paths for the circulation of Korean medicine products. Similarly, Caroline Meier zu Biesen's Chapter 6 unpacks the dynamic at play between circulation and governance in the case of artemisinin-based combination therapies (ACTs) between Asia and Africa. This author sheds light on the way by which the pharmaceutical industry fosters the global circulation of ACTs, and how this expansion and the economy it generates led to new regulatory apparatuses, and finally to the transformation of malaria control. This volume has many such stories to tell us, underlining the co-production of circulation and governance. This reciprocal affair, however, is complicated by the fact that it potentially affects the pharmaceutical object itself.

Take the case of new Ayurvedic formulations produced by a global firm established in India. New formulations are made in various batches, so that there are several possible formulations to test for one or more therapeutic indications. These early formulations enter the Department of Formulation Development, which is responsible for making a product suitable for mass production and the market. With the aim to facilitate market penetration in Europe and the United States, specialized personnel in the Department of International Regulatory Affairs intervene at this stage. They check the constituents of all formulations and may discard or keep a formulation on this basis. If a formulation contains a plant which is banned in the United Kingdom, European Union, or United States, the formulation is simply thrown out the window, irrespective of its efficacy. The formulation does not meet the target of "global compliance," and this pitfall precludes any further development (Pordié 2014). A substitute for the plant will then be found. This does not pose any major problem in the eyes of the inventors, as they think there are interchangeable plants in the classical medical texts that may enter a single formulation and which are considered to have similar effects. Conversely, if a formulation contains a plant which is part of an existing pharmacopeia in these countries, its use is strongly encouraged and the formulation kept. In-depth research will follow this selection phase, and the most active formula will be selected. Governance and the regulatory environment of Western countries play a very direct role in the innovation processes and the selection of formulations in Ayurveda. Regulation does not only inflect the trajectory of Asian medicine; it also transforms objects, weighing heavily in this case on the components of a given drug prone to global circulation.

Identical conclusions can be made about circulation and its effect on the object. Markovitz and his colleagues put it boldly: "In circulating, things, men, and notions often transform themselves. Circulation is therefore a value-loaded term which implies an incremental aspect and not the simple reproduction across space and time of already formed structures and motions" (Markovits et al. 2003, 2–3). Think about the shift in status from medicine to health supplements for a variety of Asian medicinal products between Asia and Europe (Janes 2002; Pordié 2008, 2015). As a way to circumvent the stringent regulatory environments of Europe or the United States, and to conceal their quality as "medicines," Asian medicine goods come under new labels. Once therapeutic objects in Asia, they become health supplements, functional food, or nutraceutics elsewhere. Chapter 5 by Liz P.Y. Chee on fish liver oil provides a telling example. She describes a product at the boundary of food and medicine which shifts over time from side to side according to changing regulatory environments, market orientations, and global circulation, thus blurring lines and opening possibilities. This is also shown in the contributions by Enjeong Ma, Simeng Wang and Caroline Meier zu Biesen (Chapters 2, 7, and 6, respectively), which all deal with some kinds of hybrid drugs, whether from the standpoint of regulations and product categories or from that of epistemology. Karen McNamara further demonstrates in Chapter 1 that the legal value of different categories of medicines is unstable and increasingly influenced by their circulation in the global market. In these cases, the pharmaceutical object appears to be fragile and sensitive to circumstances (Appadurai 2006). The object's conception, legal status, value, content, or therapeutic indication are potentially altered by circulation.

For the purpose of analysis, let us take health governance out of its entanglements and mutual production with circulating and transforming objects, and examine the establishment of laws and policies aiming at regulating and monitoring Asian medicine. The chapters by Céline Coderey and Arielle A. Smith (Chapters 3 and 4, respectively) show how nation-building projects and other states' agendas, in Myanmar and Singapore respectively, inform health regulation and more generally the way medicine comes to be legitimized. They both illustrate the normative power of biomedicine, and the ways expressions of biopower translate into particular modes of governance. These processes are described as political, as they usually take place in relation to central governing structures and entails remarkable transformations of therapeutic power. They are also deeply economic, as one of the chief aims is to foster the accumulation of capital. Drawing from Fassin's brilliant analysis on the genesis and transformations of medical power (2000), we also understand that health governance cannot be reduced to a mere political or economic reading. It involves the moral foundations of medicine and therapeutic power. By fixing norms for Asian medicine, these regulatory regimes convey sets of new behaviors and values, a moral code that concerns the nature of right and wrong. Céline Coderey signals these moral inflections by underscoring the disappearance of certain elements in the therapy in Myanmar, such as ritual practice and other religious components. Not only medical knowledge and practice but also "the possibility of a discourse on medicine" are thus subjected to a profound reorganization (Foucault 1963). While the contributors in this volume do not lose sight of the social, cultural, epistemological, and clinical dimensions of contemporary Asian medicine, the chapters also underscore the political character of health governance and discretely suggest its moral declination. Making this a frontline issue will open new research avenues to understand the moral process at work in health governance and regulation and, through it, in the social world.

Finally, official modes of governance cannot explain how circulating objects and people are regulated, simply because a sizable part of what and who circulate escapes institutional governance. Simeng Wang's Chapter 7 explores such routes and the many ways regulation is circumvented by practitioners. She reveals a number of strategies – which is to be understood, after Michel de Certeau (1980), as referring to subjects who claim a place of their own, a base from which relations with an exterior threat can be managed – used by people to bypass regulatory regimes. Whether it is a matter of Chinese medicine practitioners in France who use geographic mobility to escape national governance (Wang, Chapter 7 of this volume), of restocking the country of Nigeria with pharmaceuticals (Peterson 2014) or of smuggling Siddha medicine products from Southern India to Europe (Sébastia 2011), people are resourceful in finding ways to make official governance look weak. Thus, studying the shady networks and practice of Asian medicine that escape institutional governance provides a unique perspective on the role and functioning of the state at its medical and social margins (Das and Poole 2004). Following this idea, the very notions of health governance and pharmaceutical regulation, which are taken in this volume as forged by institutional authorities, should possibly be revisited and expanded, confronted as they are with multiple realities. The world is also ordered and ruled through myriads of other agents and practices that do regulate. For instance, individual agency, interrelations, and cultural differences influence regulatory processes (Brhlikova et al. 2011; Kuo 2008), just as multinational pharmaceutical companies may take on the role of regulators (Peterson 2014, 195). The gravitating forces of regulation are as much located within as outside the control of central powers (Quet et al. 2018). In this perspective, the act of regulating reflects its etymological roots, and involves all the means, people, or institutions that control or direct by a rule, a principle, a law, or a method, and, by extension, that put or maintain social activity in order. When pharmaceutical objects circulate in the real world, they go through and influence, as much as they are affected by, expanded systems of control and regulation that go far beyond the realm of institutional governance and practice. This is perhaps the subliminal message of this book.

References

Appadurai, Arjun. 2006. "The Thing Itself." *Public Culture* 18 (1): 15–21.

Brhlikova, Petra, Ian Harper, Roger Jeffery, Nabin Rawal, Madhusudhan Subedi, and M.R. Santosh. 2011. "Trust and the Regulation of Pharmaceuticals: South Asia in a Globalised World." *Globalization and Health* 7 (10). www.globalizationandhealth.com/content/7/1/10.

Buse, Kent, and Gill Walt. 2002. "The World Health Organization and Global Public-Private Health Partnerships: In search of 'Good' Global Health Governance." In *Public-Private Partnerships for Public Health*, edited by Michael Reich, 1–19. Cambridge, MA: President and Fellows of Harvard College.

Das, Veena, and Deborah Poole, eds. 2004. *Anthropology in the Margins of the State.* New Mexico: School of American Research Press.

De Certeau, Michel. 1980. *L'invention du quotidien*, 1. *Arts de faire*. Paris: Union Générale d'Editions, coll. 10/18.

Fassin, Didier. 2000. *Les enjeux politiques de la santé. Etudes sénégalaises, équatoriennes et françaises.* Paris: Karthala.

Foucault, Michel. 1963. *Naissance de la clinique*. Paris: Presses Universitaires de France.

Janes, Craig. 2002. "Buddhism, Science, and Market: The Globalisation of Tibetan Medicine." *Anthropology and Medicine* 9 (3): 267–289.

Kloos, Stephan. 2019. "Humanitarianism from Below: Sowa Rigpa, the Traditional Pharmaceutical Industry, and Global Health." *Medical Anthropology*. doi: 10.1080/01459740.2019.1587423.

Kuo, Wen-Hua. 2008. "Understanding Race at the Frontier of Pharmaceutical Regulation: An Analysis of the Racial Difference Debate at the ICH." *Law, Medicine and Ethics* 36 (3): 498–505.

Markovits, Claude, Jacques Pouchepadass, and Sanjay Subrahmanyam. 2003. "Introduction: Circulation and Society under Colonial Rule." In *Society and Circulation: Mobile People and Itinerant Cultures in South Asia, 1750–1950*, edited by Claude Markovits, Jacques Pouchepadass, and Sanjay Subrahmanyam, 1–22. New Delhi: Permanent Black.

Packard, Randall. 2016. *A History of Global Health: Interventions into the Lifes of Other Peoples*. Baltimore: Johns Hopkins University Press.

Peterson, Kerstin. 2014. *Speculative Markets: Drug Circuits and Derivative Life in Nigeria*. Durham, NC: Duke University Press.

Pordié, Laurent. 2008. "Tibetan Medicine Today: Neo-Traditionalism as an Analytical Lens and a Political Tool." In *Tibetan Medicine in the Contemporary World: Global Politics of Medical Knowledge and Practice*, edited by Laurent Pordié, 3–32. London and New York: Routledge.

Pordié, Laurent. 2014. "Pervious Drugs: Making the Pharmaceutical Object in Techno-Ayurveda." *Asian Medicine* 9 (1–2): 49–76.

Pordié, Laurent. 2015. "Hangover Free! The Social and Material Trajectories of PartySmart." *Anthropology & Medicine* 22 (1): 34–48.

Quet, Mathieu, Laurent Pordié, Audrey Bochaton, Supang Chantavanich, Niyada Kiatying-Angsulee, Marie Lamy, and Premjai Vungsiriphisal. 2018. "Regulation Multiple: Pharmaceutical Trajectories and Modes of Control in the ASEAN." *Science, Technology and Society* 23 (3): 1–19.

Saxer, Martin. 2013. *Manufacturing Tibetan Medicine: The Creation of an Industry and the Moral Economy of Tibetanness*. Oxford: Berghahn Books.

Scheid, Volker. 2002. *Chinese Medicine in Contemporary China: Plurality and Synthesis*. Durham and London: Duke University Press.

Sébastia, Brigitte. 2011. "Le passage des frontières de médecines pas très douces: prévenir l'innocuité ou préserver l'authenticité?" *Revue d'Anthropologie des Connaissances* 5 (1): 71–98.

Index